Corporate Disclosures and Financial Risk Assessment

Philipp Kissing

Corporate Disclosures and Financial Risk Assessment

A Dichotomous Data-Analytical Approach Using Multivariate Scoring Models and Scenario Techniques

Springer Gabler

Philipp Kissing
Menden, Germany

ISBN 978-3-658-12459-5 ISBN 978-3-658-12460-1 (eBook)
DOI 10.1007/978-3-658-12460-1

Library of Congress Control Number: 2015960844

Springer Gabler

Printed on acid-free paper

Springer Gabler is a brand of Springer Fachmedien Wiesbaden
Springer Fachmedien Wiesbaden is part of Springer Science+Business Media
(www.springer.com)

Table of Contents

List of Figures

List of Tables

List of Abbreviations

Act. Gr.	Actual Group
CAPM	Capital Asset Pricing Model
CEO	Chief Executive Officer
Conf. Int.	Confidence Interval
CSR	Corporate Social Responsibility
df	degrees of freedom
Dis. Sc.	Discriminant Scores
IPO	Initial Public Offering
Low B.	Lower Bound
MDA	Mahalanobis Distances Approach
NGO	Non-governmental organization
Pred. Gr.	Predicted Group
SE	Standard Error
Sig	Significance
sMDtC	Squared Mahalanobis Distance to Centroid
Std. C.	Standardized Coefficient
Std. Dev.	Standard Deviation
Unstd. C.	Unstandardized Coefficient
Up B.	Upper Bound
VIF	Variance Inflation Factor
WLA	Wilk's Lambda Approach
ZScore()	Standardized Score of ()

1 Introduction

The role of corporate disclosures in today's business and financial life is enormous. Companies that need to refinance their operating activities make efforts to reducing information asymmetries. For investors, disclosures provide an opportunity to assess potential risk and return connected to investment opportunities. The proper functioning of financial markets relies on a balanced situation of information distribution and corporate disclosures can be considered as the instrument that regulates this setting. This explains their outstanding role for the capital market mechanism.

The aim of this study is to connect both: corporate disclosures and the financial risk assessment of investors. In order to reach this objective, multivariate models are elaborated theoretically and analyzed through the application of statistical procedures. The integration of tailor-made scoring models and established scenario techniques as well as the quantitative model setting guarantee innovative results from a methodological point of view. With regard to contents, this is the first study connecting the quality of corporate disclosures, the individual decision making level of investors and the impact on overall market data. However, the ambition of this study is not limited to its academic output. The practical pertinence of the results with reference to publishing companies, investors and regulating state authorities is pointed out.

The following lines provide an outline of the structure of this work. In the second chapter, the underlying theories with reference to information asymmetries and corporate disclosures are introduced chronologically. Additionally, previous research approaches and the results of corresponding studies with reference to determinants and consequences of corporate disclosures are summarized. The conceptual framework of corporate disclosures related to voluntary elements is described in the third chapter. As a preparation for the following substudies, the research approach of this work is derived in this context.

The following two chapters are very comprehensive and provide the two quantitative substudies. The first study focuses on the relationship between the quality of corporate disclosures and their impact on overall market data, in particular unlevered beta factors as risk related indicators. Additionally, determinants that drive the disclosure quality as well as credibility factors are tested empirically. In addition to different statistical procedures, scoring models set the basic technical element of this first substudy.

In the context of the second substudy, the financial decision processes of investors with reference to corporate disclosures are researched more detailed. Through the application of scenario techniques and multivariate statistical analyses, the role of corporate disclosures on the individual level is investigated. In order to simplify the understanding, the two substudies are organized identically. The main ideas are summarized in the framework section of each study. The tested hypotheses are developed in the theoretical development parts. The research design chapter covers technical aspects as well as the elaborated quantitative models. The results are presented with reference to the hypotheses grouping and are finally discussed concerning their limitations. The sixth chapter provides a synthesis of the tested econometrical models of the two substudies and summarizes their quantitative results. Finally, this work closes with a synopsis.

2 Theoretical Foundation and Previous Research Approaches

2.1 Theoretical Foundations of Corporate Disclosures

2.1.1 Asymmetric Information and Agency Cost Theory

Akerlof's description of „Lemon Markets" (Akerlof, 1970) is not only the basis for a critical reflection of the working mechanisms of markets and their efficiency. His theory also sets the fundamental theoretical basis of corporate reporting.

Akerlof takes the used car market into consideration to explain the problem of quality uncertainty. Here, the person who intends to sell a car has extensive knowledge about the specific conditions of the car. The quality of a car varies for instance by the frequency of maintenance, the accident history and the owner's driving style. The potential buyer of a car cannot verify those features because many mechanical parts are hidden from view or in general not provable under normal circumstances.

This concept of asymmetric information can surely not only be observed at the used car market. However, the conclusion can be drawn that sellers with above-average-quality goods will tend to leave such a market. The reason is that buyers' best guesses for a given product under asymmetric information is that it is of average quality and accordingly they are only willing to pay the price of a product of known average quality.

This means from a theoretical point of view that in the worst case the interaction between quality heterogeneity and asymmetric information can lead to the disappearance of a market. Comparable market characteristics are also described by Gresham's Law which is one of the most traditional economic principles. (Giffen, 1891)

As there exist many markets with the (initial) tendency of asymmetric information and the selling side is surly not willing to accept prices that fall below the quality of the product in the long run, approaches were developed to reduce information asymmetry and its negative impact on markets.

This objective can be achieved by establishing rules that force market participants to exchange information or through voluntary information exchange. In both cases the verification of the given information are a critical success factor.

The transfer of the concept of asymmetric information to the field of corporate reporting can be described as followed. In many cases, companies are owned and managed by different (groups of) persons. While the managers are responsible for the business activities, the owners (or investors) finance the company's operations. Thanks to their activities and position, managers have a deeper knowledge about the situation of the company at the market, their own managing skills and potential operating or even financial problems of the company.

At the same time, the owners have a lower level of information about these issues. By analogy with Akerlof, information asymmetries can be observed. Within this context, the described problem is named Principle-Agent dilemma under the condition that both parties are maximizing their own benefits (Jensen and Meckling, 1976). This theorem is based on Akerlof's description of "lemon markets".

The existence of asymmetric information in the context of companies and investors causes several problems. The uncertainty of the investors about the situation of the company could lead to adverse selection which depends on the level of information asymmetries and the investors' attitudes towards uncertainty.

This ex ante problem of information asymmetry reduces the financing activities or could in the worst case lead to a breakdown of the financial supply. Irrespective of the exact result, the financial supply would deviate from Pareto efficiency as long as the investors are not affine to uncertainty. This is the first part of the so called agency costs (Jensen and Meckling, 1976). Generally spoken, for the case, that actions, which are useful for the principal yet costly for the agent, occur, these deviations from the principal's interests are called agency costs.

In addition to adverse selection, the second group of problems that arise by the principal-agent dilemma can be summarized by the term moral hazard. These problems occur when the management is already established. In general, moral hazard arises when a party is willing to take risks or to conduct activities that would not be accepted without knowing that the potential costs or burdens are taken in whole or in part by others. While the acting party has the chance to receive (partly) the positive consequences of these activities, the responsibility for a negative outcome is mainly taken by the other party.

In economic theory, moral hazard is regarded as a special case of information asymmetry because it can only occur if one party is isolated from risks and has more information about the potential consequences of the activities than the other. A further precondition is that a complete monitoring is impossible.

Applied to the situation of managers and agents on a corporate level, agents may have an incentive to decide for activities that are inappropriate from the viewpoint of the principal because their risks are limited to losing their jobs, they have a deeper knowledge about potential consequences of their acting and usually they cannot be completely monitored.

There exist several specifications of these ex post conflicts and approaches to reduce their negative impacts for instance by implementing contracts that aim to align the interests of the principal and the agent (Grenadier and Wang, 2005). These approaches propose methods like profit sharing to reduce the divergence of interests.

Apart from setting incentives to align the different interests, the reduction of information asymmetries between the involved parties leads closer to a perfect market and its benefits. Corporate disclosures offer an opportunity to reduce this asymmetry by improving the knowledge of the principals.

In order to reach this goal, the agents have to publish reliable information about the situation of the company and their activities. These disclosures can be either published on a voluntary basis or the agents are obliged to report on a regular basis. As long as the agents are interested in improving the market situation into the direction of a perfect market, disclosure obligations are not necessary.

Yet the agents may also benefit from information asymmetries and their interests to publish information are limited to that degree which makes the investors willing to finance their activities while the future risks are hidden. Consequently, compulsive reporting standards that set a minimum quality level regarding corporate disclosures have to be established.

In practice, nothing else is done by regulatory authorities. Nevertheless, voluntary reporting does not vanish. There is a large scope of activities that make voluntary reporting still relevant due to branch specific features that cannot be respected by regulatory authorities or it can be seen even used as an advantage regarding the competition for financial supply. These issues will be discussed thoroughly in a further chapter.

Disclosures provide the opportunity for corporate entities to signal superior quality and attractiveness compared to competitors at the market. This aspect may vary on the nature of reporting (voluntary or obligatory), benchmarking aspects and other factors. Analogously, the agency costs are supposed to vary on the legislative system (statutory vs. common law) or contractual issues (Jensen and Meckling, 1976).

This consideration shows that already the simple relationship between two actors in the light of asymmetric information causes demanding challenges and environmental issues increase this complexity extraordinary.

The original corporate governance model is just based on the dilemma between principal and agent as well as the control over the agency costs (Jensen and Meckling, 1976). Nevertheless, agency costs do also play a role in more situations, companies are confronted with. A vast literature on agency-theoretical explanations has been published over the last decades.

The financial leverage of a company was identified to reduce agency costs through the threat of liquidation that is connected to personal losses to managers like reputation or salaries (Williams, 1987). Moreover, the pressure to generate cash flows to make interest payments also enhances this effect (Jensen, 1986). But as already described by Jensen and Meckling (1976), the effect of leverage on the agency costs is expected to be non-monotonic.

This means that at low levels of leverage, the described positive incentives for managers reduce the agency costs yet this effect is reduced at high levels of leverage where bankruptcy become more likely. This theorem is based on a separate consideration of the agency costs that are connected to the debt and the equity. In the last case, the agency costs of debt overwhelm the agency costs of equity. Consequently, the total agency costs increase.

In addition to effects that arise from the capital structure, even more aspects that have an impact on agency costs are known in the area of corporate finance. It is discussed and empirically tested that a high dividend yield compared to corporate earnings lower agency costs (Rozeff, 1982). This effect is comparable to the effect of interest rate payments as described above. However, the raise of transaction costs of external financing reduces this positive impact and leads again to the question of the perfect capital structure of a company.

Agency cost theory has been used in many different areas like economics, accounting, organizational behavior and many more. Of course, this theory is also surrounded by controversy. Critics complain that the agency cost theory does not address clear problems (Perrow, 1986).

Its proponents argue that the agency cost theory is a revolutionary foundation for research affecting different types of organizations (Jensen, 1983). According to Eisenhardt (1989, p. 72), the valid perspective lies in the middle and agency cost theory "provides a unique, realistic, and empirically testable perspective on problems of cooperative effort."

For the purpose of this study, the presented approaches of information asymmetry and agency costs are respected in the context of corporate disclosures like the majority of research papers linked to this area also do. Nevertheless, the theoretical framework of corporate disclosures is much more complex. The analysis is not limited on the relation between corporate entities and their shareholders.

2.1.2 Stakeholder Theory

While the original corporate governance model is limited to principals and agents as well as the control over agency costs (Jensen and Meckling, 1976), there are actually much more parties involved in the scope of corporate activities. The concentration on the owners of a company and the maximization of shareholders' benefits do not respect the influence, participation and the legitimate interest of these parties. Freeman and Reed (1983) and Freeman (1984) create the first stakeholder model that broadens the limited shareholder approaches.

This theory of organizational management and business ethics addresses not merely morals and values, but rather identifies and models the stakeholder groups of corporate entities (Kochan and Rubinstein, 2000). Moreover, managerial methods are provided that support the management to recognize and integrate the interests of the different groups.

There are several definitions and differentiations of stakeholders in the academic literature. "Common to nearly all stakeholder definitions is the notion that a stakeholder is any individual or group of individuals that is the legitimate object of managerial or organizational attention." (Phillips, 2003, p. 25)

Concerning the field of corporate reporting, the various stakeholder groups are relevant from two different perspectives. On the one hand, they can be regarded as potential addressees of corporate disclosures as they are interested in the development of the company with a certain thematic priority. Secondly, they can be object of the corporate reporting.

By focusing on the stakeholder differentiation of Freeman (1984), stakeholders can be divided in internal and external ones. Based on the involvement in the company's nexus of contracts, the internal or direct stakeholders are for instance investors, employees, customers and suppliers (Baret, 2006).

External or indirect stakeholders are state authorities, local residents, competitors, media and NGOs (Baret, 2006). The existence and influence of these different stakeholder groups varies with the sphere of activity of corporations.

On the one hand, Freeman's theory receives increasing attention over the years and his approach is integrated in several research papers across different disciplines and educational books. Moreover, it is used as the leading framework in the majority of corporate social responsibility methods.

On the other hand, critics complain that the interests of the various stakeholders can hardly be balanced against each other and even a stakeholder paradox is identified (Goodpaster, 1991).

According to the argumentation of Goodpaster (1991), it seems to be illegitimate to orient corporate decisions by ethical values that go beyond strategic stakeholder considerations because the obligations of agents to principals are stronger than those of agents to third parties.

Freeman (1994) replies that the stakeholder paradox is built on wrong theses that imply separation within corporate entities and society instead of regarding business activities on a common base and in the light of ethical standards. While this disputation appears philosophically in some parts, it also shows that different interests drive the behavior and actions of the participants and conflicts can arise. With regard to the variable compensation components for managers in practice and their orientation based on financial figures in most cases, this debate is still actual.

However, this characteristic does not reduce the importance of the stakeholder theory but makes a differentiated view necessary. The evaluation of the influence of each stakeholder and its implication on the corporate attention differs on the activity field of the company and situational factors. The relevance of the various stakeholder groups with respect to the field of corporate reporting has to be ordered in order to achieve a solid theoretical basis for empirical tests of hypotheses.

As already described before, Freeman (1984) and Baret (1986) group the different stakeholders. The examination of the legitimacy of these groups within companies was subject to several studies. Phillips (2003) as well as Andriof and Waddock (2002) work primarily with the following two approaches. On the one hand, the strategic approach focuses on stakeholders' analytical aims.

On the other hand, the normative approach concentrates on the moral perspective and postulates as well as ethical orientation for the appraisement of the legitimacy of stakeholders. For the purpose of this study, the focus is on those stakeholders that are close to the corporate business activities and most responsible for the generation of economic performance as described by Hillman and Keim (2001). This restriction has to be made to enable the analysis of a sufficient large number of companies for empirical tests. Moreover, the aim of this study is to work with corporate financial figures that are essential to evaluate the risk assessment of investors.

As stakeholder theory developed to a broad field and the different stakeholder groups, their roles and influence as well as their relevance for business activities are debated, a limited stakeholder model is necessary for research purposes. Wolfe and Putler (2002) examine the stakeholder categories and propose to pay special attention to stakeholders that are involved in the company's operating cycle.

By arguing that companies are adding value through their purchasing, manufacturing and sales activities, the identification of the responsible parties in these fields can be regarded as a first approach. Husser et al. (2012) use a similar approach and argue that the regularly equated first-order stakeholder groups are consisting of suppliers, employees and customers because these stakeholders represent the economic and investment cycle of corporations. These stakeholders are also part of the list of direct stakeholders published by Baret (2006).

The usage of a limited shareholder model reduces the number of stakeholders significantly. Nevertheless, the remaining stakeholders are in the most cases not homogenous groups with a full alignment of interests. Employees and executive employees can be named as an example of persons from the same stakeholder group with partly differing interests because the executive employees fulfill to a certain degree the job of employers.

Secondly, there are only a few obligatory regulations regarding corporate disclosures with respect to suppliers, employees and customers and most information are published on a voluntary basis (Clarkson, 1995). The regulatory minimum requirements differ significantly from country to country.

Thirdly, the way to communicate with stakeholders can differ. While employees can be addressed internally and externally, the possibility to communicate with (potential) customers is more limited. All these features play an important role for the transfer of stakeholder theory to the field of corporate reporting and have to be integrated to the evaluation of the research results.

Ullmann (1985) develops a conceptual framework to explain the relationships among social disclosure and corporate economic performance. His framework is an advancement of the strategic management approach by Freeman (1984). The first dimension of his approach is "stakeholder power". It is regarded as the power of a stakeholder to influence the management and is viewed as a function of the stakeholder's degree of control over resources that are required by the corporation.

The second dimension is the corporate strategic posture towards corporate social responsibility activities. Ullmann (1985) outline two different strategic posture of the companies' key decision makers which can be named as active and passive and describe their response to social demands.

The third dimension concerns the economic situation of the company and its financial capability to social investment and its credible reporting. This approach combines the ideas of the stakeholder theory with the field of corporate reporting for the first time. However, this conceptual framework is limited to social issues.

It can be summarized that the neo-classical perspective, arguing that expenditures on social causes are reducing the shareholder's wealth and are a violation of the management's responsibility, is obsolete (Friedman, 1962). Apart from moral issues, respecting the different demands of stakeholder groups are relevant even in order to avoid diminished shareholder value through labor strikes, boycotts or lawsuits. This is the core message that can be derived from the modern stakeholder theory.

From a stakeholder theoretical perspective, the corporate performance is assessed in terms of the degree a company meets the demands of multiple stakeholders. Basically, there exist two different approaches combining the field of stakeholder theory and corporate reporting.

From the perspective of Transaction Cost Economics, companies that try to satisfy the demands of their (most important) stakeholders and signal their quality can avoid higher costs that arise from more formalized contractual compliance mechanisms (Williamson, 1985). From the resource-based perspective, companies regard the meeting of stakeholders' demands as a strategic investment. The fulfillment of these expectations can be seen as a competitive advantage that can hardly be imitated (Russo and Fouts, 1997).

Both strategic approaches are in fact a consideration of the advantages and disadvantages that arise for the company due to its activities and reporting behavior. This view is captured comprehensively and in more detail by the information cost theory that is presented in the next chapter.

2.1.3 Information Cost Theory

Information costs can be regarded in two slightly different contexts. As part of the transaction costs which arise from economic exchanges, the search and information costs are (in most cases) seen as costs that are caused by the buying party to determine the availability and the price of a product (Dahlman, 1979).

In the context of corporate reporting and for the application in this study, information costs are costs that publishing companies are confronted with. These companies have to estimate the costs that are connected to generating and publishing information (Verrechia, 2001). In both cases, asymmetric information can be identified as the basic precondition for the existence of information costs.

For the purpose of this study, the information cost theory is used as an approach that describes the corporate consideration of the costs and benefits that are connected to publishing disclosures. Cormier and Magnan (2003) examine this cost-benefit analysis and call it briefly the "communication strategy of a company". It is characterized by the balance between the benefits of publishing disclosures, the risks that could rise from stakeholder pressure and regulatory constraints.

The following two graphics provide an overview about the cost-benefit consideration and the type of different costs companies are confronted with. These underlying characteristics of the information costs theory and the influence of stakeholders as well as the impact of regulatory specifics are described below.

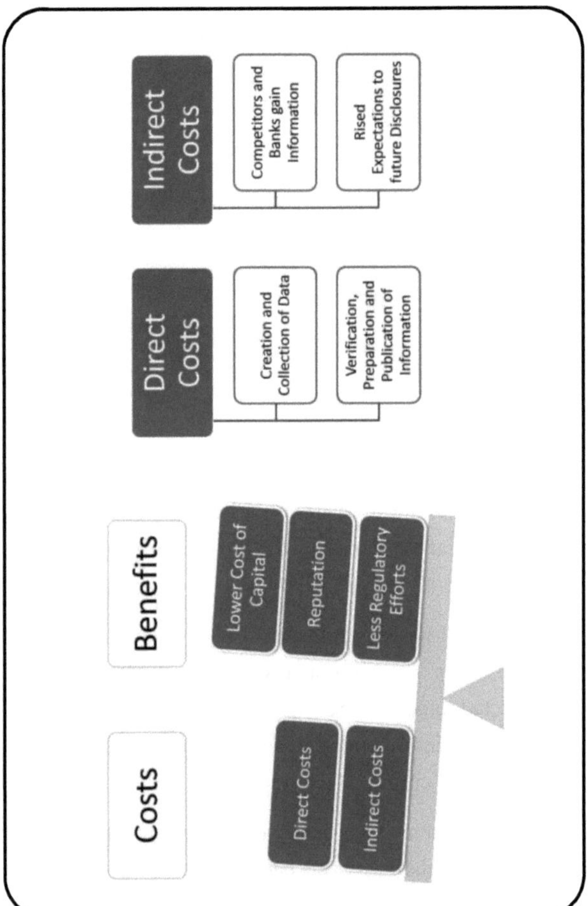

Figure 1: Cost-Benefit Analysis of Corporate Reporting and Cost Differentiation

The costs that arise from publishing disclosures can be divided initially into two parts: direct and indirect costs. The direct costs are connected to the original process of collecting and publishing the information while the indirect cost are incurred due to the behavior of market participants because of a change of their knowledge with reference to information about the publishing company.

The direct costs are driven by the creation and collection of data. Beside labor costs, the installment and service of IT resources can be named. Moreover, the implementation of a company-wide understanding for the need, the willingness and the capacity to collect (specific) corporate information has to be conducted.

These costs vary with the nature of the data (Hirst et al., 2008). Especially the collection of qualitative information is challenging because a common understanding about the exact object of measurement has to be developed at first to create reliable data (Rogers and Stocken, 2005).

Secondly, collected information has to be verified and prepared for the publication. With reference to the aim of the publication, the collected primary data also have to be compared to secondary information in order to derive necessary information and the disclosures have to be created. Lastly, the documents that are intended to be published should or have to be audited by an external party to guarantee their reliability and credibility. Especially with reference to the IT resources, some economies of scale effects are observable for the direct costs.

The indirect costs are the effects that arise from a change in the behavior of market participants or business partners of the reporting company due to the published information. These costs are also called "proprietary costs" (Entwistle, 1999).

From a theoretical point of view, these costs are actually based on game theoretical approaches. Consequently, they vary by the business field, the degree of competition and some more factors. Indirect costs can be identified especially in research-intensive areas where competitors could gain information about the publishing company which they could use for own development purposes. But also structural or organizational information can be used for benchmarking practices and potential competitive advantages of the publishing company could vanish (Hayes and Lundholm, 1996). In addition to that, banks that are involved in financing corporate projects also gain a superior knowledge about the overall corporate activities and could adopt their risk grading positively or negatively.

Lastly, companies generate by publishing comprehensive information a higher degree of expectations to their future disclosures. Issues that are reported extensively in previous years have to be addressed in a similar way in order to avoid the reports' addresses to derive interpretations from "silence" (Hollander et al., 2010). While economies of scale effects are partly observable for direct costs, indirect costs are in general not affected by them.

The listed direct and indirect costs are surely not limited to the named items. Based on specific corporate situations, there can exist more costs that have to be considered for the cost-benefit analysis (Dye, 1985).

However, these items are the most common ones that are discussed in theoretical literature and represent the theoretical idea of the information cost theory best. The potential benefits that are illustrated in the next lines, have to be regarded even more differentiated. Several studies examine different benefits and their extent. Consequently, the named benefits in the following are the most accepted ones which are empirically tested. They are in one line with the information cost theory. A more detailed empirical observation is conducted in the previous research approach section.

Companies pursue different objectives through the publication of information. Disclosures that include a wide variety of corporate information can be regarded as an instrument to improve the reputation of the company.

Beside financial figures, social and environmental issues that are important for the different stakeholders can be named as relevant reporting objects (Cormier and Magnan, 2007). The corporate image does not only play a role for the customer's decision to buy a product. It also increases the customer loyalty, reduces the uncertainty in the eyes of the distributors or retailers and helps to attract highly qualified employees (Lemmink et al., 2003).

In addition to the reputation, the publication of corporate information is connected with a reduction of the information asymmetries and external parties gain deeper information about the company's activities. From a theoretical point of view, the financing costs of the publishing company should be reduced due to less uncertainty at the investors' level (Diamond and Verrecchia, 1991). Ceteris paribus, a reduction of the financing costs increases the net value of the company (Botosan, 1997). It has to be mentioned that this argumentation works perfectly in theory and has also been proved empirically (Graham et al., 2005).

Nevertheless, several factors influence this relationship. More detailed information are provided in the research approach section. As a third category, the reduction of regulatory efforts by state authorities or other regulators can be regarded as a potential objective that is connected to the publication of information. The idea behind is that companies can use the opportunity to signal their social responsibility, their proactive management of resources and environment as well as further factors that are relevant in the specific business field (Verrecchia, 2001).

This positive behavior could reduce the necessity for a fixed level of compulsory regulative interventions (Belkaoui and Karpik, 1989). This aspect is captured explicitly by the political cost theory which is described in the following section.

The cost-benefit analysis in the context of the information cost theory can only be applied for information that are published on a voluntary basis. Figures or other issues that have to be disclosed are not part of this approach because the corporate entities are not free in their decision to publish this kind of information and a cost-benefit analysis is not conducted.

Nevertheless, there are several objects in practice that incorporate voluntary and obligatory reporting aspects. Companies that decide for a listing at a stock exchange are confronted with listing requirements that include regularly conditions with reference to reporting issues. The decision for a specific exchange and their regulations are mostly driven by economic forces and the connected disclosure obligations can only partly be seen as being entered voluntarily (Hirst et al., 2008).

Moreover, it can be discussed that competitors establish by the extent and quality of their disclosure, standards that have to be fulfilled by comparable companies due to the expectation of relevant stakeholder. Consequently, the theoretical abstract assuming, that companies are free in their decision with respect to voluntary reporting issues and can do a cost-benefit calculation, has some weaknesses in practice. This reporting framework is discussed comprehensively in the conceptual framework section of this study.

As described at the beginning of this section, information costs can also be regarded explicitly in the context of the transaction costs (Williamson, 1985). In this sphere, the information costs also play a role for the addressees when they use and evaluate the provided information.

However, this aspect of information processing and the arising costs at this process have been shown to be of lower importance (Casson, 1996). Consequently, the focus of this study lies on the defined corporate cost-benefit analysis.

Due to the increasing relevance of the publication of voluntary information especially over the last 25 years, the information cost theory is of growing importance (Verrecchia, 1990). Nevertheless, studies show that the assumption of constant information costs can be criticized and the management is confronted with incentives to withhold (proprietary costs) and incentives to publish discretionary information (agency costs) according to the corporate environment (Depoers, 2000).

This indicates that corporate reporting cannot only be regarded from the information cost theoretical perspective. However, this theory sets the fundamental explanation for the corporate decision process to disclosure voluntary information.

2.1.4 Further Specific Theories and Overview

The agency cost, the stakeholder and the information cost theory are the three fundamental theories in the field of corporate reporting. Based on this theoretical fundament, the corporate decision process to disclosure information can be analysed. Nevertheless, corporate reporting is a vivid and complex field.

Consequently, there are some theoretical approaches that play a crucial role for the comprehensive understanding of corporate activities in this area as well as for research purposes. Two of these approaches - the legitimacy theory and the political cost theory – are presented here because of their outstanding relevance especially with respect to voluntary disclosures. These specific theoretical approaches should not be understood as independent and autonomous theories. They can be interpreted as derivatives from the underlying axioms.

While the idea that corporations do not only have economic and legal obligations but also certain responsibilities to society is not new (McGuire, 1963), social and environmental reporting remains over the years a voluntary practice to a high degree (Wilmshurst and Frost, 2000). With respect to social disclosures, Hogner (1982) suggests that these publications are motivated by the corporate need to legitimise their activities.

In this sense, corporate reporting can be regarded as a kind of reaction to community expectations (Tilt, 1994). This idea is captured by the legitimacy theory which is closely related or can even been seen as a part of the stakeholder theory.

The legitimacy theory is based on the idea of the social contract. According to Shocker and Sethi (1973) any institution operates in society via a social contract. They argue that the growth and survival of each institution is based on the delivery of socially desirable ends to society and the distribution of economic, social or political benefits to groups from which they derive their power. While corporate economic performance can be quantitatively measured, social performance cannot be determined easily. Nevertheless, upcoming public expectations regarding corporate activities and the diminishing pure shareholder orientation influence companies to put more effort to social and environmental aspects.

The formal fulfillment of legal standards is surely not sufficient to influence the public opinion and public policy process. Beside annual corporate reports, advertisements and brochures also contain social responsibility information disclosure (Zeghal and Ahmed, 1990). Nevertheless, regularly published and audited annual reports have a high relevance in this area due to their binding character in contrast to advertising campaigns.

The legitimacy theory and its underlying assumption play an important role because of several reasons. Companies that aim to legitimate their activities have to publish information to stakeholders that are in most cases not internal ones. Secondly, the reporting object can differ from the regular one. Thirdly, situational factors can play a crucial role.

These situational factors have to be regarded for aspects like the risk assessment of investors according to the idea of this work. Companies like energy suppliers or banks can be named exemplary. In these areas, legitimacy theory plays a major role with regard to extraordinary events like the Fukushima accident or the financial crisis that lead to public discussions about the role and responsibility of these companies.

The proof of the legitimacy theory and its consequences have been subject to several studies. The legal framework and the role of the state are identified as relevant determinants with respect to this theory (Archel et al., 2009). Moreover, the company's business field (Guthrie and Parker, 1989) as well as the degree of pollution due to the business activities are named as important drivers for the quantity of published information (Cormier and Magnan, 2003).

The disclosure of social aspects is usually connected to the expectations of society (Orlitzky et al., 2003). In the context of the legitimacy theory, it can be generalized that corporate reporting is on the one hand a reaction to public pressure as well as expectations and on the other hand a technique to secure the corporate right to exist which is also connected to regulatory influence. However, the aspect of regulation of corporate activities is even more important for the second theoretical approach that is described in the following lines.

An emerging subfield of strategic management deals with regulations that affect corporate strategies. A higher level of sensitivity to environmental issues and the recent financial crisis have illustrated the relevance of a certain level of governmental control and strengthened regulatory incentives. Regulations always mean a limitation for corporations regarding their scope of activities and less opportunity for business activities. The optimal degree of regulatory stipulations and their implementation to rules as well as the connected monitoring process can be discussed controversially.

Nevertheless, companies try to reduce the influence to their business activities and their fiscal or other financial consequences. These ideas are captured by the political cost theory. The basic ideas for this concept were already described by Jensen and Meckling (1978) as well as Watts and Zimmermann (1978). From a theoretical point of view, the ideas and mechanism behind the legitimacy and the political cost theory are interwoven for companies that are operating in countries with democratic orders. Elected politicians are expected to represent and realize the intentions of the people. Consequently, companies should just care about voters who transfer their power with reference to corporate disclosure. However, this theoretical idea cannot be applied in practice due to different reasons. Firstly, most democratic systems are not as direct as described and parties are elected for a huge amount of different reasons.

Secondly, theme-specific working groups elaborate in most parliaments the recommendations for the decision process due to the high level of complexity of most affairs.

Thirdly, many authorities like the anti-monopoly office are partly independent from the daily democratic process but fulfill a role that is based on fundamental standards. All in all, there exist several reasons for corporations to try to reduce political costs with regard to regulatory or fiscal implications.

Belkaoui and Karpik (1989) name as a crucial motivator for companies to publish voluntary information a potential reduction of further regulations and the improvement of the corporate image in public. Additionally, the ownership structure of companies is identified as a relevant factor for political sensitivity (Ghazali, 2007). This means that beside some other factors, the ownership structure has to be regarded as a determinant that can explain the extent of voluntary disclosure with reference to the political cost theory.

As there exist regularly a relationship between the size of a company and the ownership structure, some studies focus more on this aspect with reference to their analyses (Boesso and Kumar, 2007). The policital cost theory implies two more interesting concepts. Firstly, larger companies are expected to disclosure more detailed information because they are observed closely by government agencies and are more exposed to political costs (Holthausen and Leftwich, 1983).

Therefore higher voluntary reporting standards should reduce the pressure of undesired scrutiny by governmental agencies (Buzby, 1975). Secondly, highly profitable companies have the incentive to disclosure more information with reference to the political cost theory. The reason behind is that these companies should try to demonstrate the market participants and authorities the source of their profits as well as their flawless behavior. However, the empirical results for this aspect are disputable as described more detailed in the section of the presentation of previous studies.

While the political cost theory suggest that managers are concerned with the attenuation and reduction of regulatory actions and explicit or implicit taxes, the relevance of non-governmental interest groups and others stakeholders is of increasing importance (Healy and Palepu, 2001). This shows that the political cost theory as well as the legitimacy theory should only be seen in the context of the stakeholder theory that describes the relationship of corporations and their environment in a broader sense. The graphic below illustrates the development stages of the main theories that are relevant for corporate reporting as well as their relationship whereby information asymmetries can be named as a common underlying assumption.

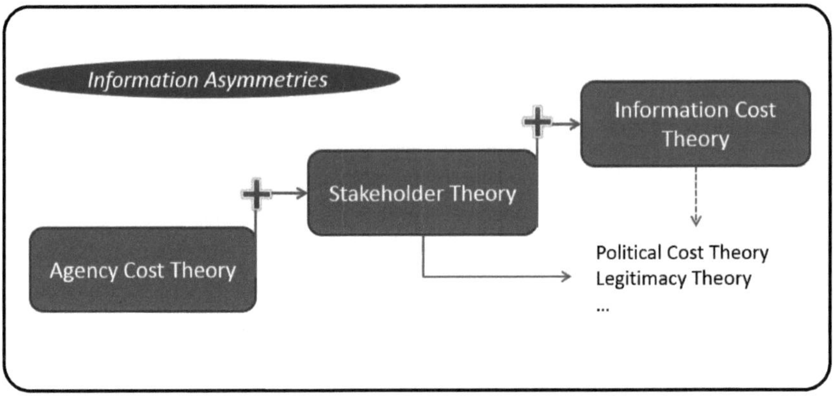

Figure 2: Overview Theoretical Foundation

Although the graphic above shows the relevant development stages of the most important theories in the field of corporate reporting, the relationship between the different theoretical approaches should be regarded more differentiated. Each theory adds new perspectives but do not necessarily replace all previous assumptions.

With respect to more specific situations, several theoretical approaches that can be regarded as derivatives of the main theories are used in the economic literature. Beside the political cost and the legitimacy theory as described in this chapter, the most popular specific theoretical approaches have already been introduced in the context of the underlying theories in the previous chapters.

The general signaling theory focuses primarily on how companies can show their quality to market participants and stakeholders. The proprietary cost theory was described in the context of direct and indirect costs with respect to the information cost theory.

2.2 Previous Research Approaches

2.2.1 Financial Effects of Corporate Disclosures

The development process of the described theories in the field of corporate disclosure was accompanied by a multitude of different research approaches. Structuring these previous studies is a challenging task. For the purpose of this study, the most famous research papers in the area of corporate disclosures are named and their main results are presented. In addition to that and with respect to the scope this study, the most important specific approaches regarding their methodology as well as their research results are mentioned in this section.

The majority of previous studies concentrates on the relationship of corporate disclosures and capital market consequences. On the one hand, this might have historical reasons because the idea of shareholder value dominated other approaches for many years. On the other hand, corporate activities are regularly driven by the idea of the pursuit of profit and this leading thought does not exclude corporate reporting.

Additionally, the availability and verifiability of financial figures are potential explanations for the concentration on this research object. Nevertheless, especially recent studies deal more and more with multidimensional questions with respect to the corporate environment, regulatory implications and specific corporate situations.

Regarding the potential financial effects of corporate reporting, the studies are divided into two different groups. It can be distinguished between research approaches that test the implications of corporate disclosures for the capital costs of companies and studies that examine the consequences for corporate securities.

As already mentioned in the context of the introduction of the information cost theory, the main theoretical idea behind the relationship of corporate disclosure and the capital cost is the following. Based on more information that are submitted through corporate publications, investors can make more accurate estimates of the parameters of an asset's return or payoff distribution (Clarkson et al., 1996). Consequently, there exists less uncertainty if information regarding the parameters are disclosed.

Botosan (1997) measured the association between voluntary disclosure and the cost of equity by regression analyses. He identified for companies that attract a low analyst following that a greater amount of disclosure is related to significant lower cost of equity. Botosan and Plumlee (2002) re-examined this associations based on deviant data and a different analytical method. They found that the cost of equity decreases in the qualitative level of the annual report but increases in the level of timely disclosures. As a potential explanation of the second effect which is contrary to theory, they propose that - in consistency with managers' claims - greater timely disclosures may lead to a higher stock price volatility that increases the cost of equity.

However, Cormier et al. (2009) describe in their study that a quantitative increase of voluntary reports and a higher precision level of voluntary disclosures in the field of social and human capital disclosure reduce the volatility of the share price as well as increase Tobin's Q ratio. This already shows that the research results are partly not consistent due to the application of different methods, the use of unequal data sets or the observation of divergent time periods or markets.

Due to the heterogeneity of the different studies, Leuz and Verrecchia (2000) examine in their paper German companies that switch for the first time from a national to an international reporting regime. Consequently, these companies had to commit themselves to an increased level of disclosure. Under this circumstances, the effect of the increase of the corporate disclosure could be isolated well and Leuz and Verrecchia (2000) found that the bid-ask spread as well as the trading volume that were regarded as proxies for the information asymmetry component of the cost of capital, developed in the predicted direction.

In addition to this result, Hail (2002) proofed on a sample consisting of Swiss companies a significant influence of voluntary disclosures on the capital cost. The cost advantage of capital of the most forthcoming companies in the sample was around 2 %. In contrast to this result, Déjean and Martinez (2009) could not conclude in their study on French companies that a higher extent and quality of disclosure lead necessarily to a reduction of the cost of equity. However, their analysis is limited to environmental reporting issues.

Based on the published academic literature and empirical results, Botosan (2006) reviewed the controversial question whether companies can reduce their cost of capital due to a higher level of disclosure. She highlights in her paper that the first challenge is already the measurement of the cost of equity. As there regularly does not exist a directly observable measure, the costs of equity have to be estimated. Due to different methods that can be applied, the estimations vary.

Additionally, famous multifactor models like the CAPM have estimation problems. Given this difficulty, a multitude of divergent proxies are used beside those multifactor models that hamper the comparability across different research finding even more. Nevertheless, Botosan (2006) draw the conclusion that the theoretical hypothesis that greater disclosure lowers the cost of equity is supported by many studies but additional research is necessary to explain the anomalous results in the literature.

Beside studies that examine the influence of corporate disclosure on the cost of capital, another large group of research papers investigates the impact on corporate securities. Diamond and Verrecchia (1991) studied the causes and consequences of a security's liquidity. They found that especially large companies can benefit from positive financial effects that are related to voluntary disclosure.

More precisely, they proofed that there is a link between the liquidity of the market for a firm's security and its reporting behavior because large institutional traders are attracted particularly through a significant reduction of information asymmetries. The study of Healy et al. (1999) shows comparable results for their sample consisting of 97 companies. The increase of the disclosure rating is accompanied by increases in the companies' stock return as well as the stock liquidity. Based on their study that focused on the German capital market, Leuz and Verrecchia (2000) found evidence that a higher level of corporate reporting leads to lower bid-ask spreads and more liquidity of corporate securities as already mentioned before.

While several studies elaborate this positive relationship between the increase of disclosure and liquidity figures, other papers examine whether companies try to use this effect. Watts and Zimmermann (1990) discussed that companies publish private information in situation in which they try to increase their financial market values. Verrecchia (1983) proofed that also silence can be interpreted as a negative signal. With regard to environmental issues, Carroll (1979) named different types of industrial ecological behaviors of companies already more than 30 years ago.

Some recent studies concentrate on the more specific environmental reporting area. While Déjean and Martinez (2009) could not find evidence for French companies that disclosing more information improve corporate financial figures, Plumlee et al. (2010) derive from their empirical work on companies from the USA divergent results. They measured a positive relation between voluntary environmental disclosure quality and the value of a company.

More precisely, they found that higher quality environmental disclosures are positively associated with expectations of future cash flows and emphasize that analyzing separate components of the firm value is crucial. Their result is consistent with the work of Cormier and Magnan (2007). Additionally, Husser and Evraert-Bardinet (2014) could show that investors consider environmental information to measure the performance of a corporation and social disclosure aspects impact simultaneously the short and long-term performance of the corporation.

The results for the empirical examinations of the relationship between the social disclosure quality and the financial performance figures of the reporting company vary strongly. While Patten (1991) could not measure that a higher extent of social reporting influences corporate financial figures, Cochran and Wood (1984) found evidence that a positive relationship exist.

Regarding the amount of inconsistent research results in this area, Orlitzky et al. (2003) conducted a meta-analysis of 52 earlier published studies to provide a comprehensive review and examination of the previous work. They found evidence that corporate efforts regarding social and environmental responsibility are likely to pay off. They identified a high correlation between accounting-based measures of the corporate financial performance and the social performance.

The results for market-based indicators were weaker. However, some companies could try to cover and justify an insufficient financial performance by emphasizing their social and environmental activities (Jahdi and Acikdilli, 2009). This can be named as one reason why the credibility of corporate responsibility reporting activities suffers (Chen and Bouvain, 2009). Transparent information as well as the verification of social and environmental data is essential to provide reliable information (Piechocki, 2004). This leads to a controversial debate about the ideal strategic behavior of companies to meet social and ecological expectations (Lougee and Wallace, 2008).

The choice to disclose information on a voluntary basis can in most cases be seen as a strategic decision in the light of the introduced theories. By using a sample consisting of oil and gas companies from Canada, Cormier and Magnan (2002) studied their financial reporting behavior. Their work provides evidence that the reported incomes are smoothed. While there are factors that reduce this effect like the firm size or USlistings that imply more regulatory-monitoring, this strategic reporting behavior might be anticipated by the addressees. In the worst case, this would avoid the positive financial effects of voluntary corporate reporting.

However, the work of Kasznik (1999) shows the limits of the influence of corporate choices in practice and supports the idea of the positive association between voluntary disclosure and financial effects due to a reduction of information asymmetries. He studied the management's discretion over accounting choices in the context of voluntary reporting and found evidence that managers use for their predictions reliable methods to reduce and avoid forecasting errors because they fear costly legal actions by shareholders and a loss of credibility.

It can be summarized that the majority of previous studies identified a measurable link between corporate reporting and financial figures. However, this effect varies strongly by the reporting object, the corporate environment and company-specific characteristics.

2.2.2 Non-financial Effects of Corporate Disclosures

Corporate disclosures are not only used to communicate shareholder-relevant information. In the context of the stakeholder theory, there exist many potential addressees as described in the theory section. On the one hand, comprehensive corporate reporting can improve the image of companies which plays a crucial role for the customers' decision to buy a product. Additionally, it increases the customer loyalty, reduces the uncertainty of distributors or retailers and helps to attract highly qualified employees (Lemmink et al., 2003).

All these features should finally lead to an improvement of the financial figures of the company as long as the company is profit oriented. This idea is captured through the previous chapter. On the other hand, reporting is also part of the manifestation of corporate responsibility and its justification.

However, there are only a few studies that examine non-financial effects of corporate reporting activities. This might have two main reasons. Firstly, the measurement of non-financial effects is difficult. While reliable financial data are widely available, non-financial effects can only be measured indirectly in most cases. Secondly, apart from the idea of legitimacy that supports the securing of corporate existence, most elements of the presented theories are in line with the approach that corporate reporting activities going beyond mandatory elements are finally reflected in financial figures. Nevertheless, in the following lines there are provided some examples for non-financial effects of corporate disclosures.

The improvement of the corporate image can be named as one intended effect of corporate reporting. Mercer (2005) argued in his publication that the aim of building up a superior level of reputation can only be achieved by high quality voluntary reporting over a long period of time. In contrast to this study, Aerts et al. (2006) found evidence that intra-industry imitation effects exist with respect to environmental reporting behavior.

This means that there is an industry related converging tendency of corporate disclosure over time which would mitigate the positive effects of corporate reporting due to a reduction of uniqueness and credibility. Nevertheless, Aerts et al. (2007) also identified a positive association between web-based corporate performance disclosure and analyst behavior and renewed the findings of Lang and Lundholm (1993).

Beside analyst ratings, a very few studies documented further non-financial effects of corporate reports. The work of Perrini (2005) can be named exemplary as an article that provides an overview of non-financial disclosures on corporate social responsibility and potential consequences. In the context of the publication of Ittner and Larcker (1998) who examined whether non-financial measures can be regarded as indicators of financial performance, these kinds of studies are also relevant from a pure economic perspective.

It can be summarized that the number of published papers that examined non-financial effects of corporate disclosure is rather small. The reasons are named above. Several studies identified a positive relationship between the reporting quality and the stakeholder power or the media exposure (Cormier and Magnan, 2007).

However, these factors should be regarded as drivers that influence the reporting quality instead of effects arising from corporate disclosures. The following chapter provides an overview about the empirical tested determinants of corporate disclosures.

2.2.3 Quality Determinants of Corporate Disclosures

Structuring previous research results with respect to factors that impact the quality and extent of corporate disclosures is a challenging task. On the one hand, a large number and variety of different approaches exist. On the other hand, the used samples consist of companies with different sectorial and country-specific backgrounds. In order to provide a clearly arranged overview that is consistent with the conceptual framework of corporate reporting which is presented in a further chapter below, the tested determinants are divided into two groups.

Company-specific conditions characterize the first of these two groups. Aspects like the financial situation, the corporate management and compensation system, the resources, the direct and indirect cost of publication as well as the degree of information asymmetries play a crucial role here. The second group consists of determinants that concern the corporate environment. Regulatory stipulations, the industrial sector or the impact of external stakeholders can be named exemplary as drivers that characterize this area.

Based on the analysis of environmental disclosures of French companies, Cormier and Magnan (2003) identified the firm size as one crucial determinant of the environmental reporting strategy of companies. Moreover, they name proprietary costs as well as the media visibility of corporations as further determinants that impact the environmental reporting behavior. Déjean and Martinez (2009) worked also with a sample of French companies and focused on environmental reporting. They found that the environmental disclosure quality is related to the firm size, the financial leverage as well as financial analysts following the company.

While the influence of several determinants varies over different studies, the firm size can be named as the most reliable driver of the quality of environmental disclosure (Patten, 1991). This determinant was tested successfully for companies from different backgrounds (Cormier et al. 2005). However, the firm size is not measured consistently. Most studies use the market capitalization but some focus on book values, sales figures or other more specific values. This limits the comparability of the results among the different studies.

Additionally, Cormier et al. (2005) found evidence for large German companies that the ownership structure, the fixed asset age and the previous reporting behavior affect the environmental disclosure quality. With respect to the scope of this study, their findings regarding perceived risk and corporate disclosures are relevant. They argue that a higher level of risk which is captured by volatility, forces companies to provide more information. In order to measure the risk, they use the CAPM beta factors of the corporate stocks. Although they identified a positive association between the beta factors and the environmental disclosure quality, their findings are limited because they ignored the corporate capital structure as an important parameter of the beta factors (Modigliani and Miller, 1958; Leland and Toft, 1996).

Several studies applied leverage as an explanatory variable with respect to the voluntary disclosure quality (Botosan, 1997). Highly levered companies are more likely to disclose comprehensive information in order to reduce legal risks (Watts and Zimmerman, 1990).

Comparable determinants were discovered through the analysis of web-based corporate disclosure. Factors like the firm size, the corporate strategic focus and the institutional governance can be named exemplarily (Aerts et al., 2007). With regard to disclosures of research and development activities, Jones (2007) found that companies with lower book-to-market ratios tend to publish more information. This can be explained by the low degree of information that is provided through regular basic financial statements regarding market values and the corporate efforts can be regarded as one reason for the disproportionately high value assessment by market participants. With respect to the publication of sensible information, strategic optimized behavioral pattern can be identified. Lakhal (2005) proofed that the cost of information influences significantly the voluntary disclosure quality at the French financial market.

Rogers and Stocken (2005) studied the ability of market participants to interpret published management earnings forecast and found a comparable relationship between the cost of information and the voluntary reporting quality for companies from the USA.

Nagar et al. (2003) focused on the relationship between the stock price-based compensations and the disclosure strategy of managers. From a theoretical point of view, the share in ownership should reduce the agency problem in the context of corporate disclosures. Nagar et al. (2003) found evidence that the corporate disclosures of their sample are positively related to the proportion of stock-based compensation and the value of the hold shares of the CEO. Consequently, the compensation plan of the management can be regarded as a company-specific determinant that influences the extent and quality of corporate disclosures. This relationship is consistent with several studies that identified that the firm's governance is a crucial determinant regarding the reporting strategy.

Cormier et al. (2012) considered web-based voluntary disclosure under governance attributes. They identified the board of directors as an important actor in firm's reporting strategies and highlight the relevance of efficient governance on the voluntary disclosure quality. Nevertheless, Cheng and Lo (2006) examined whether insiders make use of the disclosure policies and the timing of voluntary disclosures in order to maximize their own trading profits. They identified a positive correlation between the publications of bad news forecasts and CEOs' stock purchasing activities. Mangers tend to time their trades by buying fewer shares beforehand and more after the disclosure of bad information. However, this relationship could not be confirmed for the management's selling activities.

In addition to the described company-specific determinants, several research papers that examine the impact of corporate environmental factors exist. Brown and Higgins (2001) found evidence, based on their international sample across 12 countries, that the relevance of prospect reporting varies significantly across nations which affects the quality of earnings forecast. Aerts et al. (2007) found comparable differences between the European and North American market with respect to corporate performance disclosure.

On the one hand, the legal environmental that implies a different level of regulatory stipulations and control can be regarded as an actor that influences the corporate reporting behavior. Cormier et al. (2012) elaborated the legal system as a credibility factor of voluntary disclosures. Webb et al. (2008) proofed the impact of the legal environment on voluntary disclosure and confirmed the interdependence.

On the other hand, the legal system is not the only factor that characterizes the corporate environment. Guthrie and Parker (1989) as well as Cormier and Magnan (2003) identified the industry membership as one of several drivers that determines the corporate reporting management. This determinant represents in fact several other potential drivers like the degree of competition and its impact on corporate reporting or branch-specific expectations for instance concerning social issues (Orlitzky et al., 2003). All in all, the majority of the successfully tested determinants in this area have been derived from stakeholder and legitimacy theory.

It can be summarized that a large variety of potential determinants, regarding individual corporate attributes as well as the corporate environment that influence the reporting quality, have already been tested. The results are not always consistent but provide indication for corporate incentives to disclose information. The determinants may vary over the course of time due to changing market conditions but their awareness is essential to interpret corporate disclosures.

2.2.4 Academic Voids

Although a considerable amount of research papers in the area of corporate disclosure was created over the last decades, academic voids exist with respect to the applied methodologies, the used data sets as well as the defined research objects. In general, the transfer of gained knowledge based on country specific samples has to be scrutinized critically due to cultural diversity and different legal requirements.

The stakeholder theory and its inherent idea that several parties have a legitimate interest in the corporate activities is widely accepted nowadays and companies publish different kinds of specific reports in addition to the obligatory financial statements. Nevertheless, the primary focus has been on capital market consequences up to now. Further impacts of these disclosures on their addressees have rarely been object to research papers.

Additionally, the continuous change of market conditions creates a challenge for corporate entities but at the same time new research tasks. Upcoming technical opportunities like web-based publications of corporate disclosures set new options with regard to the timeliness of the published information, the interactivity of different types of stakeholders and provability.

Future research papers in this area deal with a largely unknown field that opens many potential research approaches but they are also faced with several problems regarding the reliable record of information especially for retrospective methods.

Moreover, the vast majority of previous studies focused on analyzing corporate disclosures over short time periods. Long-term observations could enrich the academic research from two different perspectives. On the one hand, the positive effects of corporate disclosure are closely connected to the credibility of the company. Consequently, corporate reputation can solely be built up by consistent corporate disclosures over the course of time (Mercer, 2005). A significant part of the consequences of corporate reporting can only be observed and interpreted based on multi period samples.

On the other hand, market conditions are changing continuously as already described before. As corporate reporting can be regarded as a reciprocal process that is affected by changing environments, the consequences of corporate disclosures and the adaption of corporate reporting behavior due to these changes, long term analyses could deliver many insights and pertinent facts about the reporting nature of companies.

However, the realization of multi period analyses is connected to an enormous work load regarding the evaluation of the corporate reports if detailed content analyses are conducted. This explains the lack of this kind of studies.

The growing number of standardized reporting indices that provide information about companies with regard to different reporting issues can be regarded as an upcoming option to work on large samples. Of course, several limitations especially with reference to the evaluation process have to be considered here as well.

The relevance of sustainability with regard to economic growth and the interrelations of the responsible use of resources, environmental issues as well as the interests of human beings and their work forces play today a crucial role for corporate entities. Nevertheless, the publication of sustainability disclosures has only been object to a limited number of studies.

Moreover, the consequences of these disclosures are rarely examined. From a methodological point of view, the consideration of published information along the entire value chain and its implications for the risk assessment of investors or comparable issues have not been researched.

There is a huge lack of studies that examine the impact of comprehensive disclosures on market data that incorporate the investors' risk assessment as well as the influence of corporate disclosures on the individual human decision level to buy or sell a financial security.

Lastly, extraordinary macroeconomic events like the subprime crisis as well as exceptional company-specific situations like IPOs or increasing regulatory pressure might impact the (voluntary) reporting behavior of corporations. These specific situations can also be seen as the basis of further research.

3 Conceptual Framework and Research Approach

3.1 Conceptual Framework of Voluntary Disclosures

The basis for the research approach of this study as well as for the derivation of the research questions is an in-depth analysis of the situation publishing corporate entities as well as financial investors are confronted with.

This analysis can only be conducted by clarifying the conceptual framework in the field of corporate disclosures with a focus on voluntary reporting elements. Several basic assumptions that are fundamental for this study are introduced in this chapter. The graphic below illustrates the conceptual framework of voluntary disclosures which is verbally elucidated in the following lines.

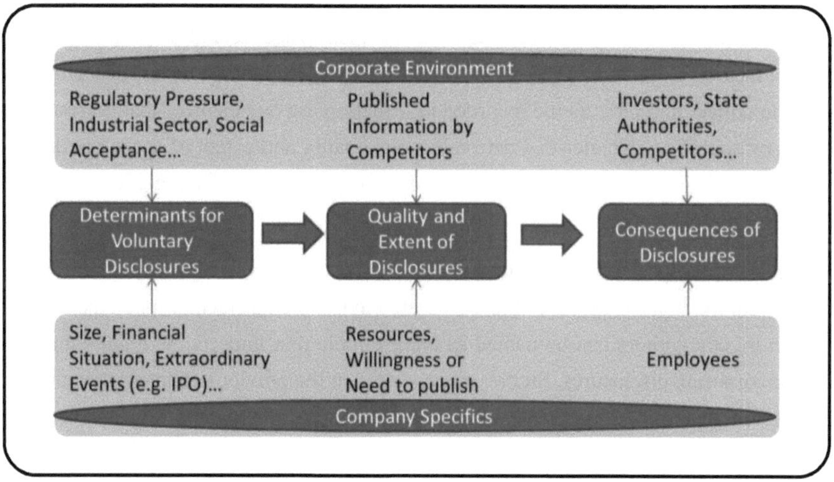

Figure 3: Conceptual Framework of Voluntary Disclosures

On the one hand, the illustration of the conceptual framework of voluntary disclosures provides an overview of the complex situation publishing companies as well as their addressees are confronted with. On the other hand, it gives a first impression about the variety of potential determinants that might impact the reporting behavior of companies.

In general, voluntary corporate reporting activities are characterized by two different groups of parameters: company specific attributes and the corporate environment. The influence of the elements of these two groups is notable along the whole reporting process. Potential determinants that drive the initial decision to disclose information can be sorted in one of these two groups.

Company specifics like the size which regularly represents the relevance of the corporation for the society, the financial situation or extraordinary events like an IPO that is characterized by a high level of information asymmetries can be named exemplarily. Regulatory pressure, the branch of industry or the social acceptance of the corporation and its activities are potential determinants from the corporate environment that force companies to disclose voluntary information.

With regard to the quality and the extent of the published information, potential explanatory variables can be divided again in company specific and corporate environmental factors. Corporate resources with respect to the human and technical features in order to collect and process the recorded information are one element that represents a company specific parameter that determines the quality and extent of corporate disclosures. The willingness or need to publish information can partly be regarded as a company specific element as well.

However, this aspect is often influenced by external factors like public opinion pressure or comparable. With regard to corporate environmental factors, published information by competitors can be named as one example that impacts the quality and extent of corporate disclosures. Studies as presented in the previous chapter have already discovered that imitation effects with reference to corporate reporting are notable. From a practical point of view, benchmarking activities are also applied in the corporate reporting area.

Lastly, potential consequences of the published disclosures can be grouped into company specific and corporate environmental ones as well. On the one hand, internal addressees like employees could adopt their behavior with respect to motivational issues or planning activities for future career opportunities based on the published information if the disclosure contains relevant topics. The consequences that can arise from activities of external parties are more complex and multifaceted. Depending on the published information and the expectations of investors with reference to the financial market in general, investors can adopt their investment management which impacts the price of traded shares and securities as well as the success of current and future corporate issuing activities in order to raise new capital.

Additionally, state authorities could proof critically the published information to evaluate the entrepreneurial practice and check the necessity of further regulatory interventions or the provision of public incentives to direct corporate activities.

Moreover, competitors could use the disclosed information to gain knowledge about the publishing company with regard to information sensitive areas like research and development activities or just with reference to best practice approaches in other fields. All these potential consequences that are driven by the adaption of strategic behavior can already be seen as determinants that influence the voluntary reporting decision of the publishing company because of the freedom of choice to disclose information in the area of voluntary corporate reporting.

Consequently, the conceptual framework of voluntary reporting is characterized by a reciprocal cycle whereby determinants as well as the consequences of the reporting activities can be divided into internal and external factors. The variety and changing influence of these factors characterize the complex situation publishing companies are confronted with and have to be considered in order to obtain reliable research results.

3.2 Underlying Assumptions

The relationship between corporate disclosures and figures that can be observed at financial markets can only be examined if it is clarified how market participants deal with available information.

While early studies focused solely on random walks and martingale models to describe the attributes of stock prices implicating stochastic price processes (Cowles and Jones, 1937), Fama (1970) developed the efficient market-hypothesis.

Although the efficient market-hypothesis is discussed controversially and behavioral finance economists proved irrational inefficiencies, three major levels of this hypothesis are defined: weak, semi-strong and strong efficiency (Fama, 1970).

While the weak level claims that market prices reflect all information that were available in the past, the semi-strong definition states that the prices additionally change instantly in order to reflect all new public information.

The strong efficient market-hypothesis claims that prices instantly reflect all information including hidden information. Most empirical investigations provide evidence in favour of the weak and semi-strong form (Khan, 1986). The application of this knowledge in the field of corporate reporting leads to the conclusion that stock prices should reflect published information.

However, processing the disclosed information regularly needs a certain time depending on the extent of the disclosed report and complexity of the circumstances as a whole. These specific characteristics should be considered for research approaches in the field of corporate reporting.

Based on the work of Markowitz (1959) on diversification and modern portfolio theory, Sharpe (1964) developed the famous capital asset pricing model to determine the adequate rate of return of an asset.

Although the model is discussed critically over years and the arbitrage pricing model by Ross (1976) has some advantages like the missing precondition of market balance, the CAPM is widely applied in the financial industry e.g. in order to calculate the asset's sensitivity to non-diversifiable risk.

The adjustment of these so-called beta factors in order to incorporate firm specific conditions like the capital structure is necessary to examine the impact of corporate disclosures on these factors. This allows drawing conclusions how the risk assessment of investors is affected by corporate reports.

Dhaliwal et al. (2011) have already examined the relationship of corporate social activities and the cost of equity and found that companies with superior social responsibility performance can reduce their cost of equity capital.

However, their approach is limited to social issues and do not consider sustainability reporting elements. Moreover, they used the arithmetic mean of the results of three different models that estimates the implied cost of equity capital as a proxy of the companies' costs of equity.

Research papers that aim to measure the impact of corporate disclosures on financial market values often use the modified model of Ohlson (1995). This model provides a conceptual approach for financial accounting research in general.

The basic model develops the idea that a corporate value analysis can be conducted by calculating the sum of the book value and the present value of the expected abnormal earnings. While the abnormal earnings satisfy an autoregressive process, dividend payments are treated as paid out of the book value instead of the current earnings.

The modified model of Ohlson is used to conduct linear regressions in the area of sustainable development to proof potential reporting consequences (Plumlee et al., 2010). By using the market value of a traded share as the dependent variable, the share book value and the earnings as well as further independent variables like the reporting score can be used to test the impact of corporate disclosure.

The knowledge of the introduced theories as well as their potential practical consequences are presupposed in the context of this work. The exact methodical approaches and their variations in the scope of this study as well as the specific underlying theoretical assumptions are explained in detail in the corresponding parts of the two substudies.

3.3 Derivation of Research Approach

Almost all kinds of markets are characterized by different degrees of information asymmetries. This study focuses primarily on voluntary corporate disclosures elements that are used to reduce information asymmetries between corporate entities and their stakeholders.

Firstly, the impact of the reduction of information asymmetries through voluntary corporate reporting elements on risk representing market data is tested. Secondly, the effect of information from corporate disclosures on financial decision making processes of individual investors is analyzed on the microeconomic level.

As social and environmental disclosures have partly a substituting impact on the reduction of information asymmetries (Cormier et al., 2011), the first substudy uses sustainability reporting elements as an integrative approach.

The application of analytical grids allows the structural analysis of corporate disclosures that contain information regarding sustainability issues.

The analytical grids are developed to fulfill two main conditions. On the one hand, economic, social and environmental information are evaluated. This collection allows a comprehensive analysis that considers financial and non-financial parameters. On the other hand, this study is not limited on specific corporate areas but takes disclosed information along the whole value-added chain into consideration.

The second substudy uses scenario techniques in order to research the decision making processes of financial investors more detailed. As both substudies elaborate insights of the relationship between corporate disclosures and financial investors either on the market or the individual level and their results are combined in the synthesis section of this work, the overall leading research question of this research project is the following:

Do corporate disclosures affect the risk assessment of financial investors?

In order to answer this question on the market level through statistical procedures, the mostly qualitative information that are published by corporations have to be transferred to quantitative data.

To this end, the applied analytical grids provide different sub-scores that represent the reporting extent and quality of each company in the different reporting areas.

The structural analysis of the corporate disclosure as well as its scoring is conducted to cover the relevant operational risks of the corporate activities. Corporate financial figures are widely available and taken from the most famous financial databases.

Some data have to be adapted to incorporate specific financial risks like the capital structure. These data are used for several statistical tests to examine the impact of voluntary reporting activities on the risk assessment of investors.

Additionally, companies are grouped according to disclosure patterns and their main reporting objects with reference to different stakeholder groups are examined. Determinants that can explain various quality levels of corporate disclosures are tested, too.

In order to research the individual decision level of investors with reference to information from corporate disclosures, scenario techniques are applied and the numerical results are recorded to test the developed quantitative models.

Consequently, this thesis shall lead to a better understanding of corporate disclosures and their consequences for the financial risk assessment. The remainder of this work consists of two substudies that focus on the macroeconomic as well as the microeconomic level of this question.

4 Corporate Disclosures of Mid-Caps and Market Risk Indicators

4.1 Framework

The purpose of this first substudy is to investigate corporate sustainability disclosures, determinants that drive their information quality as well as consequences on the market risk assessment of financial investors on a sample of mid-caps. The analyzed corporations were MDAX-listed.

The framework of this substudy can be defined as follows: As direct and indirect costs are incurred for the publication of information and managers are expected to behave rationally, the disclosure of information about the sustainable development should also have a beneficial aspect for corporations. Investors care about future-related information and their assessment should play an important role when they value a corporation.

However, the credibility of reporting activities of corporations especially with reference to the publication of voluntary information plays a crucial role. Based on these assumptions, the relationship of sustainability disclosures and the risk assessment of investors are tested empirically and disclosure elements that can improve corporate financial figures are elaborated. In this context, paid dividends are introduced as a potential factor to verify the adequateness of disclosed sustainability information.

"Sustainability is the [emerging] doctrine that economic growth and development must take place (...). It follows that environmental protection and economic development are complementary rather than antagonistic processes" (Ruckelshaus, 1989). Although this statement was published more than two decades ago, sustainability is more topical than ever before. Corporations are aware of the relevance of sustainable development and align executive remuneration schemes with long-term business sustainability and performance (Pass, 2003). However, the understanding and the definition and scope of corporate sustainability differ.

As introduced in the first parts of this thesis, sustainability is not limited to its environmental and social perspective in the context of this substudy. The aim of this research approach is to analyze the relationship between corporate disclosures, their determinants and the assessment by financial investors comprehensively. As investors are usually not oriented towards the common good but interested in their own benefits, the economic perspective of sustainability is also taken into consideration.

However, these three dimensions are not necessarily in conflict but rather complement one another in future-related perspectives. The production of environmental friendly products can be a competitive advantage due to an increasing environmental awareness of customers or regulatory impacts. Emissions along the corporate value chain can increase the basis of taxations or comparable costs. The trading in emission rights can be named as an example. As corporations have two options to avoid additional costs which are relocating their emission intensive activities abroad or introducing environmental improvements, investors are interested in the corporate decision making process due to its financial consequences.

Comparable aspects can be discussed with reference to employees' rights at corporations or their suppliers. However, in pursuing mitigation and avoidance strategies, corporations are likely to follow short term objectives and may risk their reputation and credibility (Gray and Balmer, 1998). Many more examples could be named to elucidate the strong link between the environmental, social and economic dimension of sustainability. The existence of cartel fines and comparable penalties should only be mentioned here.

In fact, corporate sustainability is more than the compliance with laws and regulations. Nevertheless, the design and process of sustainable activities as well as the corresponding reporting is mainly realized on a voluntary basis. In highly competitive markets, sustainability activities could even be regarded as a level where comparative advantages can be initiated and realized.

In order to conduct a comprehensive examination of the sustainability reporting activities, environmental, social and economic elements along the corporate business cycle are analyzed. It can be expected that corporations that publish informative disclosure with reference to their purchasing, producing and distributing activities can reduce uncertainty premia of financial investors. This should be reflected in sensitive risk indicators especially in challenging market environments. To this end, 50 MDAX-listed corporations were analyzed by focusing on the difficult trading year 2011. Additionally, this approach adds value by researching medium-sized enterprises in contrast to the majority of studies that concentrate on leading indices.

The remainder of this substudy is organized as follows. In section 2, the theoretical development of the tested hypotheses is presented. The applied research methodology and its assumptions are introduced in the third section. Afterwards, the empirical research findings are shown. The fifth section concludes with a discussion of the findings and its limitation as well as a perspective for future research approaches in this field.

4.2 Theoretical Development

In this chapter, the tested hypotheses of this study are derived from widely accepted economic theories. Moreover, important previous studies that are related to these hypotheses are shortly introduced. The assumptions that underlie the test of determinants that drive the disclosure quality are presented in the first part. The second part deals with the consequences of sustainability reporting activities.

4.2.1 Theses Development of Disclosure Determinants

The information that is analyzed for this study was disclosed on a voluntary basis. Larger companies are expected to publish more detailed information which leads to higher corporate disclosures scores according to the analytical grids that were applied. This assumption can be justified by several reasons.

Firstly, larger companies have in relation to their size positive economies of scale effects regarding the direct costs of disclosing information. Direct costs are driven by the creation and collection of data due to an input of labor costs, the installment of IT etc. These costs vary with the nature of the data (Hirst et al., 2008). Especially for the collection of qualitative information, the exact object of measurement has to be developed at first to create reliable data (Rogers and Stocken, 2005).

Secondly, larger companies are more sensitive to public observation and exposed to higher political pressure which increases the amount of disclosed information (Watts and Zimmerman, 1990). This can be explained by their high relevance for society and potential high costs of failures (Skinner 1995). Moreover, the publication of voluntary disclosures can be regarded as a strategy to reduce further regulatory influence (Belkaoui and Karpik, 1989).

Many previous studies in the area of corporate disclosures tested empirically the firm size as a determinant that drives the reporting quality and quantity. Cormier and Magnan (2003) identified the firm size as one driver that influences the environmental reporting strategy of companies significantly. Comparable results were identified by Déjean and Martinez (2009) who focused on environmental disclosures. According to Patten (1991), the firm size can even be named as the most reliable driver with reference to the quality of environmental disclosures.

Moreover, a positive relationship between size and disclosure quality could not only be observed for environmental disclosures. Aerts et al. (2007) found a significant influence of the firm size on web-based corporate disclosures. Clarkson et al. (1999) identified a significant impact on the corporate social performance. Additionally, this determinant was tested successfully for companies from different backgrounds (Cormier et al. 2005). It can be summarized that the firm size is established as a crucial determinant that influences positively the information quality of corporate disclosures according to previous studies.

However, the firm size was not measured consistently. Some studies used the market capitalization and others focused on book values, sales figures or other even more specific values. In accordance to the majority of the previous studies and to avoid short-term market influences, the book value of total assets is used for this research approach. Due to technical reasons, the natural logarithm was calculated to perform the multivariate regression analysis as done by most researchers like Cormier and Magnan (2003). In this way, the first hypothesis can be formulated as follows:

H1: Large companies publish more comprehensive disclosures than smaller companies.

Secondly, the leverage effect due to financial debt is taken into consideration. While the academic literature regarding the leverage effect in financial markets is very rich and many studies research different capital structures and their consequence, it could initially be proved that financing decisions are not relevant in perfect markets as the firm value is determined by corporate assets and not by splitting the cash flows (Modigliani and Miller, 1958).

However, perfect financial markets cannot be observed in practice because of different reasons. Taxes affect corporate decisions with respect to the capital structure; costs of bankruptcy and financial distress do exist and the issue of equity as well as agreements of debt contracts consume resources. Additionally, information asymmetries as mentioned in the theoretical part of this thesis influence the relation between a company and its debt and equity holders.

To this end, the relevance of financial leverage to corporations and their stakeholders in practice cannot be ignored and there may also be a link between corporate disclosures and financing activities. Fama and French (1992) suggest that the leverage effect impacts the cost of equity and the firms' communication strategy. They argue that companies in poor financial condition may be unable to withstand the initial negative consequences of the publication of bad news. Consequently, these corporations might avoid the publication of comprehensive information although this is needed to gain benefits from more extensive disclosure (Aerts et al., 2007).

More precisely, it can also be argued that different capital structures and the corresponding divergent importance of debt holders versus equity holders may influence the reporting strategy of corporations. If corporations make agreements of debt contracts, sensitive information are shared between the company and the financing party. In this case, the public disclosure of sensible data is not necessary or at least its relevance is reduced due to the lower importance of shareholders. Based on this argument, a higher degree of debt would speak in favor of a lower motivation to publish voluntary information as direct and indirect costs arise.

With reference to the theory of proprietary costs as described in the theoretical section of this thesis, Cormier and Magnan (2003) tested empirically a negative relationship between a firm's leverage and the environmental reporting performance. Their results are consistent with prior findings (McGuire et al., 1988). Consequently, a negative impact of a firm's leverage on the disclosure quality is expected. Analogously, the second hypothesis of this substudy is:

H2: The corporate leverage is associated negatively with the disclosure quality.

Different types of market-to-book ratios are established as intuitive measurements to make rough estimates about the valuation of corporations and their activities. Based on the market-to-book ratio of equity, it can be observed how much more or less compared to its book value, market participants are willing to pay for the equity of listed companies. From the theoretical perspective in the context of information asymmetries, investors should be willing to pay exactly the book value if they behave risk-neutral and have no knowledge about the corporation and their activities.

However, corporations are obliged by law to publish financial statements. Nevertheless, these statements are mostly backward-oriented and investors care about the future development of the corporation. This development can be judged in a more comprehensive way based on future-oriented reporting issues. Risk, R&D or sustainability disclosures can be regarded as crucial elements for this assessment. Plumlee et al. (2010) measured a positive relation between the voluntary environmental disclosure quality and the value of a company.

Based on the situation that investors are indeed confronted with information asymmetries, care about the future situation and conduct risk analyses of the situation of corporations, interdependency between a corporate disclosure and the market-to-book value of equity is likely to exist.

As long as corporations behave rationally within the context of the asymmetric information and the information cost theory and the market valuation of equity is a negative mispricing, they should disclose information so that investors can be convinced that the corporation is worth more than just its book value. In order to reach this objective, companies have to publish more and comprehensive disclosures than others.

Lang and Lundholm (2000) proofed that corporations tend to publish more voluntary information in critical situations like capital increases in order to overcome the described difficulty. As credibility is costly and reputation can only be built up by consistent high quality reporting over many years (Mercer, 2005) or by sending costly signals as tested in the second part of this substudy, companies with high market-to-book ratios of equity are expected to publish more comprehensive disclosures than others.

Previous studies dealt with this and comparable ratios in different ways. Some studies focused on underpriced corporations and their reporting behavior. Jones (2007) proofed successfully that companies with lower book-to-market ratios tend to publish more information which vice versa means that companies with higher market-to-book ratios should disclose more comprehensive reports. Cormier and Magnan (2007) used the market-to-book premium for their environmental reporting study. Following the argumentation before by focusing on the long-term objectives of corporate reporting, the third hypothesis can be formulated as follows:

H3: Companies with higher market-to-book-value of equity ratios publish more comprehensive disclosures.

Additionally, the industry membership of the corporations may also influence the quality of corporate disclosures. According to Guthrie and Parker (1989) as well as Cormier and Magnan (2003), the operating fields impact the reporting management of companies. This factor may represent in fact several other potential drivers like the degree of competition or branch-specific expectations concerning social and environmental issues (Orlitzky et al., 2003). In accordance with prior findings, the operating fields of the corporations in the sample are treated as control variables.

4.2.2 Theses Development of Disclosure Consequences

In order to research the consequences of corporate disclosures by focusing on financial performance indicators, the information cost theory has to be considered. Based on this approach, corporations publish information in the hope of receiving economic benefits in return. Consequently, there should be a link between the extensity and quality of corporate disclosures and specific financial market data.

Actually, two main theoretical concepts that explain this relationship exist. Firstly, corporate disclosures reduce the information asymmetries and lead to a higher level of transparency. This leads to an important approximation of a relevant dimension of perfect markets which is generally speaking preferable for market participants. It can be followed that this approximation would finally lead to lower transaction costs. Investors are likely willing to increase their demand for corporate stocks or bonds of corporations that publish comprehensive disclosures. As the prices of traded financial products simple follow the effective rule of the balance of supply and demand, price changes should be noticeable.

Hereby, corporations that publish comprehensive disclosures are expected to reduce their refinancing costs due to an increasing demand for their stocks and bonds. This conclusion of an existing negative correlation between the cost of equity and the quantity of disclosed information with a focus on published voluntary non-financial information has already been established (Botosan, 1997).

Secondly, the role of financial analysts has to be taken into consideration. As long as financial markets are not perfect, they should assist market participants to reduce the information asymmetries through analyzing available or partly hidden information. Of course, this is the ideal interpretation of their role and their interests are driven by several more aspects in practice.

Nevertheless, it could be shown that financial analysts influence the extent to which consideration will be given to CSR information (Dejean and Martinez, 2009). If analysts are willing to invest time to follow and analyze them, companies are said to be more attractive. This is because well-explained and argued opinions expressed by experts give investors reliable information which they are then more willing to use. And corporate disclosures are a very important source of information. The integration of additional non-financial information into analysts' recommendations may also reduce the overall opaqueness at financial markets.

Indeed, a fall in equity costs has been observed if a growing number of analysts issue recommendations about corporate stocks and information asymmetries were reduced (Botosan, 1997). A recent study by Dhaliwal et al. (2012) showed that the precision and reliability of financial analysts' predictions could be improved whenever they were able to rely on information from CSR reports.

These studies support the before mentioned argumentation. Moreover, previous tests showed a connection between the stock volatility and the analyst coverage (Bhushan, 1989). Comparable results could also be confirmed in recent studies (Lang et al., 2003). They suggest that the demand for analysts' services is larger if companies have a higher exposure to financial risk and analysts connect social performance to stock returns (Luo et al., 2015).

Following this argumentation, a connection between corporate disclosures and corporate financial figures can be expected. As sustainability disclosures reduce information asymmetries and support analysts to build up their recommendation on multi-faceted sources of information, more comprehensive reports are expected to improve financial risk indicators. This is formulated in the following hypothesis:

H4: Sustainability disclosures are associated positively with financial risk indicators.

The situation between stakeholders and corporations represented by their management is characterized by information asymmetries (Akerlof, 1970). As managers have an interest to maximize their own benefits which are not necessarily in line with the interests of the owners, the so called Principle-Agent dilemma appears (Jensen and Meckling, 1976). Corporate disclosures can be used as a method to reduce these asymmetries. Nevertheless, companies and their management behave rational and strategic in this context. Based on the information cost theory, the costs and benefits of disclosing information are taken into consideration (Verrechia, 2001).

According to Baret (2006), a general focus with reference to the reporting activities is on direct stakeholders like investors. However, these investors are aware of the described situation and the strategic decision making process of corporations. Consequently, the assessment of corporate disclosures is not only limited to their extent and quality. Additional aspects also influence the evaluation of corporate reporting activities.

Previous studies showed that reports' addresses even derive interpretations from "silence" (Hollander et al., 2010). If issues were reported extensively in previous disclosures, corporations should address them in a similar way in order to avoid misperceptions or interpretations due to a lack of information. Moreover, Webb et al. (2008) identified that the legal environment has a significant influence on the disclosure of voluntary information. Cormier et al. (2012) elaborated the legal system as a credibility factor of voluntary disclosures.

Consequently, information asymmetries and the benefit-cost-model motivate the addressees of corporate disclosures to assess the reported information critically or even to interpret "silence". By focusing on sustainability disclosures and financial investors, companies should offer a tool so that investors can verify the published information and so avoid the cheap talk phenomenon (Crawford, 1998).

The verification is especially relevant if the published information are of qualitative nature and future-oriented. In this context, the dividend yield can be regarded as a potential verification factor to overcome this dilemma. Sustainability disclosures mostly relate to the overall strategic direction and long-term development of corporations. As paying dividends is connected to a cash outflow and not for free for the corporations and cutting dividends is generally seen as an alarming signal for all stakeholders and managers try to avoid it (Miller and Rock, 1985), dividends could be considered as a verification indicator.

Consequently, companies that are publishing comprehensive sustainability disclosures and are able to pay high dividends should be rated well by financial investors. The adequateness of disclosed information about the sustainable situation of the corporation is supported by a high dividend yield. This assumption does not work for corporations that do not follow a dividend policy. However, dividend policies were extensively researched and are applied in practice (John and Williams, 1985; Allen et al., 2000). Based on this argumentation, this is the fifth hypothesis:

H5: The influence of the disclosure quality on the risk indicators is stronger if the publishing companies pay above-average dividends.

4.3 Research Design

The present study includes two main stages. Firstly, structural analyses of corporate sustainability disclosures were conducted. Secondly, statistical tests of the findings in combination with further financial data were performed.

4.3.1 Data Collection

Data were collected through the application of structural content analyses on the published sustainability disclosures of all MDAX-listed companies. To this end, the corporate annual reports of the business year 2010 were analyzed with a focus on sustainability issues. If published, specific reports like sustainability or CSR disclosures were analyzed as well. All disclosures that were published in German or English were considered. Specific disclosures that are not published on a yearly basis were not analyzed if they had been published in 2008 or before.

The latest disclosure that was taken into consideration was published in March 2011. Of the 50 companies that were MDAX listed at the beginning of 2011, 36 % published additional disclosures. However, some companies published extracts of the annual reports as CSR or sustainability disclosures and even one company called the German version of their annual report "Geschäfts- und Nachhaltigkeitsbericht" (business and sustainability report). Based on the analytical method, double counting of disclosed information was avoided.

An overview of the measurement of all variables that are used for this substudy are enclosed (see appendix A). The relevant Datastream company codes of the analyzed corporations are listed there as well (see appendix B).

4.3.2 Sample Description

For this study, all companies that were MDAX listed at January 3rd, 2011 according to the official publication of the market data and analytic department of the German stock exchange (Deutsche Boerse) were analyzed.

These 50 companies were grouped in three major sectors of activity: Finance, Industry, Service & Trade. The following table shows the analyzed corporations and the assigned sectors of activity.

Table 1: Analyzed Companies

Sector of Activity	Name of the Company	Number of Companies (total: 50)
Finance	Aareal Bank, Deutsche Euroshop, Deutsche Wohnen, Gagfah, Hannover Re, IVG Immobilien	6
Industry	Aurubis, Bilfinger Berger, Brenntag, Continental, Demag Cranes, EADS, ElringKlinger, Fuchs Petrolub, GEA Group, Gerresheimer, Gildemeister, Heidelberder Druckmaschinen, HochTief, Kloeckner & Co, Krones, Lanxess, Leonie, MTU, Rational, Rheinmetall, Salzgitter, SGL Carbon, Stada, Suedzucker, Symrise, Tognum, Vossloh, Wacker Chemie	28
Service & Trade	Axel Springer, BayWa, Celesio, Douglas Holding, Fielmann, Fraport, Hamburger Hafen, Hugo Boss, Kabel Deutschland, Praktiker, ProSiebenSat.1 Media, Puma, Rhoen Klinikum, Sky Deutschland, TUI, Wincor Nixdorf	16

Different parameters that represent the characteristics of corporations were used for the statistical analysis in the context of this study. The most important attributes are summarized in the table below. A definition of each parameter is provided in the attached variable measurement table.

Table 2: Company Characteristics

	Obs.	**Mean**	**Stand. Dev.**	**Min**	**Max**
Assets (in million of euros)	50	7382,39	13533,99	301,9	78937
Beta (levered)	49*	,987246797	,3760139327	,2668907	1,7666035
Beta (unlevered)	49*	,659132062	,3027607606	,0999145	1,2620325
Leverage (%)	50	43,7818	31,12258	,99	204,62**
MV/BV of Equity	50	1,712701973	1,1869200266	-1,3600785**	6,1524498
Dividend Yield (%)	50	1,8816	2,35432	0	13,83

*one company delisted in 2011 ** one company with negative book value of equity

4.3.3 Structural Analysis Method

On the one hand, the disclosures analysis is structured to provide reliable quantitative information about the corporate disclosures. On the other hand, its set-up is organized to take the relevant sustainable factors of the corporate value-adding process into consideration.

Structural analysis methods as initiated by Barthes (1981) and implemented by Demazière and Dubar (1997) support the avoidance of three obstacles as described by Husser et al. (2012): Human beings tend to select illustrative excerpts. This leads to a loss of content. Secondly, a lack of interpretation can occur if persons follow their restorative attitude. Thirdly, subjectivity can be named as a crucial problem. The approach of this study to use an analytical framework supports the avoidance of these obstacles and the provision and analysis of reliable data.

The quality of the corporate disclosures in the sample was evaluated by a structural content analysis: For this purpose, three analytical grids (see appendices C, D and E) were used. These grids were developed by focusing on the three most important stakeholder categories which are suppliers, employees and customers (Hillman and Keim, 2001). These stakeholder groups reflect the economic and investment cycle along the corporate value-adding process best.

Each grid is related to one of these stakeholder groups. It consists of two general questions that evaluate the positioning and relevance of the corresponding stakeholder group within the reporting framework of the corporation. Additionally, further aspects that cover the three dimensions of sustainability (economic, social and environmental dimension) are integrated in these grids. These dimensions are also called pillars of sustainability (Husser et al., 2012). For each dimension, two additional aspects are proven.

This approach allows conducting differentiated analyses on the stakeholder related aspects as well as on dimension related aspects. According to the information quality of each analytical aspect, the grids provide three possible levels of information which lead to higher scores for the provision of comprehensive information and vice versa. Consequently, subscores that relate to a stakeholder or a sustainability dimension can be derived and interpreted separately. The total sustainability score for each corporation can be determined by summing the scores of the single grids.

Moreover, the analytical grids were developed to incorporate the main recommendations of the Global Reporting Initiative (GRI) as well as to integrate approaches of former research studies (Cormier et al., 2005). Consequently, different indicators of sustainability are taken into consideration whereby the provision of precise quantitative information was scored above declarative statements.

4.3.4 Statistical Analyses and Models

In order to gain an overview about the sample and to describe its population, standard statistical procedures were conducted on the results of the disclosure analysis. Moreover, explorative statistical methods were applied. A principal component analysis was conducted on the stakeholder and sustainability dimensions (9 variables). For this extraction the varimax rotation with Kaiser normalization was used.

Additionally, hierarchical classifications of the extracted components as well as of stakeholder related subscores were conducted to group the companies of the sample. As a clustering method, the squared Euclidean distances were calculated. The results of these analyses are provided in the first part of the research findings section.

However, the main research aspects of this study were tested by applying multivariate regression analyses. A first linear regression was conducted to test the determinants that drive the quality of the corporate disclosures. For this purpose, the total sustainability scores of the companies that were determined through the application of the analytical grids, were treated as the dependent variable.

The normal distribution of the disclosures scores (DS) was tested successfully beforehand (Kolmogorov-Smirnov test, $p=0{,}827$). The potential determinants were treated as independent variables within the linear regression set-up. These variables consisted of corporate data of the year 2010 and were defined as follows.

For the firm size (FS) the total book assets were determined. For technical reasons and in accordance with the majority of previous approaches, the natural logarithm of the assets was calculated and used. The total debt of total capital was defined as the leverage (LE). The equity valuation (EV) was defined as the market value of equity divided by the book value of equity. The market value of equity was calculated by multiplying the share prices by the number of ordinary shares in issue at the end of 2010.

The three different operating fields (Finance (FI), Industry (IN), Service&Trade (ST)) of the companies are coded binary and treated as dummy variables. For mathematical reasons the third dummy variable is dispensable within the context of a linear multiple regression. Nevertheless, in order to ease the understanding of the tested model, all dummy variables are stated below. In the context of this analysis, the operating fields are control variables.

The determinants of the quality of the corporate disclosures were tested by a multivariate linear regression model which was defined as followed:

$$DS_{Total} = \beta_{10} + \beta_{11}FS + \beta_{12}LE + \beta_{13}EV + \beta_{14}FI + \beta_{15}IN + \beta_{16}ST + \varepsilon_{1t}. \quad (1)$$

The consequences of the disclosed information were tested by additional regression analyses. The different subscores like the customer-related score (DSCUST) were used for detailed examinations. The tests focused on the disclosure impact on unlevered beta factors of the corporations. As the usage of industry beta factors is common, the above mentioned three operating fields were tested separately. The German stock exchange (Deutsche Boerse) provides betas for the listed corporations. Missing betas due to delisting were determined through regressions of the daily returns of the stock on the daily MDAX returns. All betas are based on figures of the complete year 2011 that followed the reporting period the disclosures are related to. Consequently, the time lag is minimized.

Based on the beta factors, unlevered betas were determined in order to incorporate the different capital structures of the companies. For this purpose, several steps were necessary: The model of Modigliani and Miller (Modigliani and Miller, 1963) which includes corporate taxes, defines the required rate of return on equity (r_{EQ}^{UL}) for a levered company as followed by including the corporate tax rate (t):

$$r_{EQ}^{LV} = r_{EQ}^{UL} + (1 - t)\frac{Debt}{Equity}(r_{EQ}^{UL} - i). \quad (2)$$

The capital asset pricing model (CAPM) as presented by Sharpe (1964) defines the levered and un-levered rate on equity as followed:

$$r_{EQ}^{LV} = i + (r_m - i)\beta_{LV}. \tag{3}$$

$$r_{EQ}^{UL} = i + (r_m - i)\beta_{UL}. \tag{4}$$

The market return (r_m) as well as the riskless interest rate (i) are not needed after plugging equation (3) and (4) in equation (2) and solving:

$$i + (r_m - i)\beta_{LV} = i + (r_m - i)\beta_{UL} + (1 - t)\frac{Debt}{Equity}(r_m - i)\beta_{UL}. \tag{5}$$

$$\beta_{UL} = \frac{\beta_{LV}}{1+(1-t)\frac{Debt}{Equity}}. \tag{6}$$

The unlevered betas (β_{UL}) were calculated using formula (6). This approach is also conform to existing accounting standards (compare IAS 36.A19). Consequently, the corporate tax rates had to be determined as the last missing variable of the equation. In Germany, the corporate tax rate consists of the uniform national corporate tax rate plus solidarity surcharge and a local tax rate. For the local tax rates the average taxes of the year 2011 provided by the Federal Statistical Office (Statistisches Bundesamt) were used. Potential tax benefits due to cross border activities were not taken into consideration in order to work with a reliable, constant total tax rate that is not exposed to biases due to one-time items.

As EADS (since 2013: Airbus) is established in the legal form of a Dutch corporation (naamloze vennootschap), the average Dutch tax rate reported by the Central Agency for Statistics (Centraal Bureau voor de Statistiek) was used in this case.

The impact of the disclosures on the unlevered betas was tested by using the following simple model exemplary formulated based on customer-related subscores and corporations from the service & trade industry:

$$\beta_{UL} = \beta_{20} + \beta_{21} DS_{CUST} + \varepsilon_{2t}. \tag{7}$$

Finally, to improve the model, the dividend yield paid by each corporation was integrated as a moderator within the regression concept. Based on the individual dividend yields per share as a percentage of the share prices by the end of 2010, the companies were divided into three groups (dividend yield: low ≤1,5%, medium 1,51%-2,5% and high ≥2,5%).

The moderator was defined as the mathematical product of the standardized dividend yield group (ZDIV) and the standardized stakeholder related disclosure score. These tests were conducted on the global as well as on the stakeholder related subscores.

4.4 Research Findings

In this chapter, the research findings are presented. In the first section, the results of the descriptive and explorative statistics are shown and illustrated graphically. In the second part, the empirical results of the hypotheses testing are presented. Further interesting aspects that were identified during the disclosure analysis are mentioned in the third section of this chapter.

4.4.1 Descriptive and Explorative Statistics

The application of the analytical grids shows promising results. The total disclosure scores of the 50 corporations of the sample vary between 68 (minimum) and 176 (maximum). The arithmetic mean of the scores is 117,6 and the standard deviation is 27,20144. The 25th percentile is 99, the median is 116 and the 75th percentile is 135. The boxplot below illustrates distribution of the total disclosure scores:

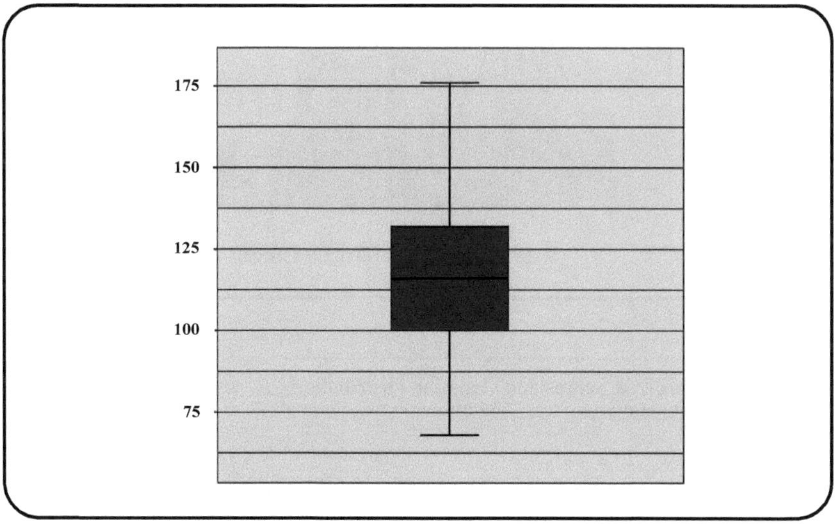

Figure 4: Boxplot of Total Disclosure Score

Finally, the distribution was empirically tested. The sample was standardized and compared with a standard normal distribution. To this end, the Kolmogorov–Smirnov test was run and confirmed that the scores are normally distributed (Kolmogorov-Smirnov Z = 0,827). Consequently, inductive statistics like the multivariate regression analysis to test determinants are applicable using parametric tests. The standardized scores as well as the Gaussian distribution are illustrated by the graphic below.

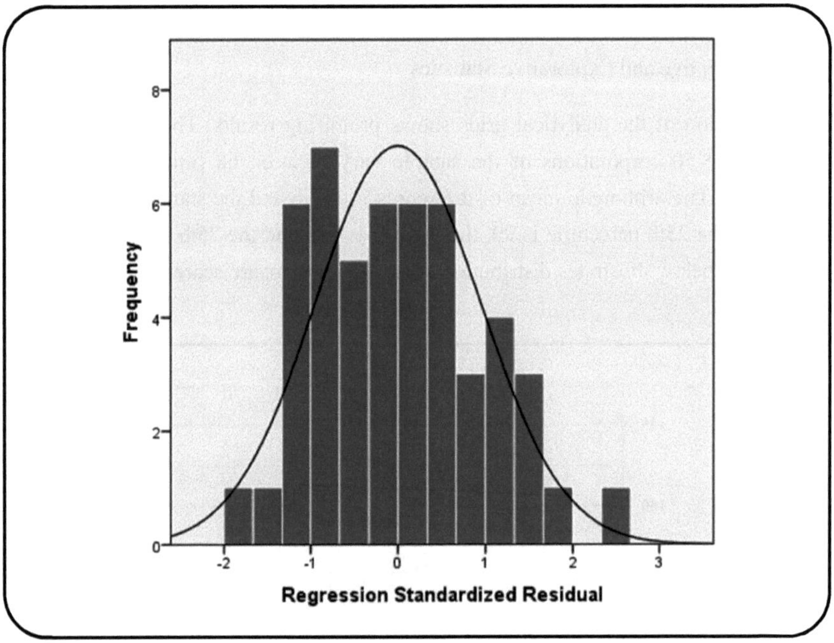

Figure 5: Standardized Scores and Gaussian Distribution

The following table illustrates the result of the total disclosure score with reference to the three different operating fields of the researched companies.

Table 3: Descriptive Statistics of Scores in Different Operating Fields

Operating Field	Obs.	Mean	Std. Dev.	Min	Max
Finance	6	81,3333	15,31883	68	104
Industry	28	122,5714	22,65920	84	176
Service&Trade	16	122,5000	28,57971	72	168

Additional and more detailed analyses were conducted on different subscores. The following table shows the stakeholder- and the dimension-related subscores. These scores are related either to one of the three stakeholder groups or to the reporting dimension.

Table 4: Descriptive Statistics of Subscores

	Obs.	Mean	Std. Dev.	Min	Max
Stakeholder-related Scores					
Supplier	50	29,20	14,25711	8	60
Employee	50	46,32	15,18517	8	72
Customer	50	42,08	12,10134	16	72
Dimension-related Scores					
Economic	50	30,64	7,89151	10	54
Environmental	50	25,20	10,44324	6	50
Social	50	27,60	9,03508	10	50

The means of both subscores show interesting differences. Particularly, the supplier-related subscore seems to be clearly lower than the other stakeholder-related subscores. Consequently, further analyses were run. Firstly, normality tests were used to proof the distribution of the stakeholder-related subscores. As there is no significant deviation from a normal distribution according to the Kolmogorov–Smirnov test, the means of the stakeholder-related scores are compared and examined for statistical significant differences.

While the employee- and customer-related subscores do not significantly differ from another, the t-test shows highly significant differences of the mean of the supplier-related subscores to the mean of the employee- ($t = -8,491$; $p = 0,000$) as well as the customer-related subscores ($t = -6,388$; $p = 0,000$). The 95% confidence interval of the difference of the supplier-related subscore to the employee-related subscore is (-21,1718) to (-13,0682). The 95% confidence interval of the difference to the customer-related subscores is (-16,9318) to (-8,8282).

It can be summarized that the supplier-related subscores are lower compared to the other subscores although the analytical grids do not ex ante imply lower scores. Nevertheless the positive correlation between the supplier-related subscores and the total disclosure scores are comparable to the other subscores. Consequently, a potential bias that might arise due to the conception of the applied analytical grids can be rejected. Nevertheless, the intensity of disclosing supplier-related issues is lower compared to other stakeholder groups in this sample which is further discussed in the conclusion part.

Comparable examinations were conducted regarding the dimension-related subscores. Again, no significant deviation from a normal distribution could be identified applying the Kolmogorov-Smirnov test for the three dimensions. Consequently, parametric tests were used to analyze statistical significant differences of the means.

While the environmental- and social-related subscores do not differ significantly from another, the t-test shows highly significant differences of the mean of the economic-related subscores to the mean of the environmental- ($t = 4,874$; $p = 0,000$) as well as the social-related subscores ($t = 2,724$; $p = 0,009$). In this case, the 95% confidence interval of the difference of the economic-related subscore to the environmental-related subscore is (3,1973) to (7,6827). The 95% confidence interval of the difference to the social-related subscores is (0,7973) to (5,2827).

With respect to the dimension-related subscores it can be summarized that the economic-related subscores are significantly higher than the others. Again, the analytical grids do not ex ante imply higher scores and the positive correlation between the different dimension-related subscores and the total disclosure scores are comparable. Potential reasons for this deviation are discussed in the corresponding section of this work.

Moreover, explorative analyses were conducted. A principle component analysis on the stakeholder and sustainability dimensions was run. The 9 variables (Supplier-Economic, Supplier-Environmental, Supplier-Social, Employee-Economic, Employee-Environmental, Employee-Social, Customer-Economic, Customer-Environmental,

Customer-Social) could be reduced to 3 factors that explain 66,895 % of the total variance. The varimax rotation with Kaiser normalization was used for this extraction.

Factors 1 consists of social and environmental related variables, factor 2 consists of customer related variables and factor 3 is more economic related. As customer related variables created an own factor, further analyses were conducted with a focus on this stakeholder group as described in the following chapter. However, clustering based on these three components did not deliver interpretable results and was not successful.

As the clustering was not successful, a hierarchical classification was conducted on the 3 stakeholder related disclosure scores (supplier, employee, customer). In order to group the corporations, the squared Euclidean distances were calculated. The corporations could be ranked in 4 groups. However, the average profiles did only deliver very limited insights as two groups consist only of 3 corporations. Due to the focus on the essentials, the detailed results are not described here.

4.4.2 Hypotheses Testing

The different determinants that might drive the disclosure quality of companies were tested within the concept of a multiple linear regression analysis. As the companies were divided into three operating fields, two of them were coded as dummies while the third group is taken into consideration implicitly. The results of the multivariate regression are summarized in the following table.

Table 5: Results of Multiple Regression Analysis (Quality Determinants)

	Hypothesis	Direction of Action (expected)	T	Significance
Firm Size	H1	+	2,224	0,031
Leverage	H2	-	-1,384	0,173
Equity Valuation	H3	+	2,075	0,044
Operating Fields			3,392-3,551	0,001
F		6,726		0,000
R^2		0,433		
Adj. R^2		0,369		
N		50		

The fundamental requirements of a multiple linear regression were fulfilled and the residual analysis provides reliable results. The total Variance Inflation Factors (VIF) is below (12). Additionally, the collinearity statistics deliver VIF below (3) for each independent variable of the regression. The explanatory capacity of this approach with 43,3 % is satisfactory. The regression is significant (F = 6,726; p = 0,000).

As expected, the firm size influenced the disclosure quality positively. The impact of this determinant could be proofed significantly (p = 0,031). This means that larger companies tend to publish disclosures of superior quality. A higher corporate leverage was expected to reduce the disclosure quality. Indeed, a higher leverage reduced the disclosure score (t = -1,384). However, this determinant was not significant within the tested regression framework (p = 0,173).

The equity valuations had a positive influence on the disclosure quality of the corporations. This determinant was again significant (p = 0,044). Lastly, the three operating fields (Finance, Industry, Service&Trade) were tested as control variables that had a significant impact on the disclosure quality (p = 0,001). Based on the results of this analysis, hypotheses 1 and 3 were accepted. Hypothesis 2 was rejected.

The impact of the published disclosures on financial figures was tested by additional regression analyses. Primarily, simple linear regressions were used to proof the impact of the total disclosure score as well as the different stakeholder related disclosure subscores on the unlevered betas. Significant results were identified for the service & trade companies of the sample (n = 16). Within the context of a simple liner regression, the beta factor as the dependent variable could significantly be explained by the customer-related subscore (F = 24,506; p = 0,000). Higher disclosure scores lead to lower corporate beta factors (t = -4,950; p = 0,000) The explanatory capacity of this approach with 63,6 % (adj. R Square = 0,610) is good. Hypothesis 4 is confirmed with reference to the service & trade companies.

However, this model could even be improved by including the dividend yield as a moderator variable. To this end, the mathematical product of the standardized customer related disclosure score and the standardized dividend group were used as the moderator variable. The extended regression analysis had an increased explanatory power with 0,739 % (adj. R Square = 0,699). The beta factors were explained significantly with this model (F = 18,426; p = 0,000).

All independent variables of this approach were significant and a higher customer related score (t = -6,067; p = 0,000) leads again to lower corporate betas whereby this effect is stronger if the reporting corporations pay higher dividend yields (t = 2,264; p = 0,041). Consequently, the dividend yield as a moderator drives the inverse relationship between the customer-related disclosure scores and the corporate betas.

The corresponding residual analysis guarantees reliable results. Hypothesis 5 was tested successfully. However, the impact of corporate disclosure elements on beta factors is limited to the 16 service & trade companies of the sample and could not be confirmed significantly on the remaining companies that were analyzed.

The main results of this substudy are illustrated graphically below. This figure shows the tested determinants that drive the disclosure quality, the corresponding hypotheses as formulated before as well as their statistical significances. The reporting consequences and the relevant hypotheses as well as their statistical significance are shown in the right part of the figure.

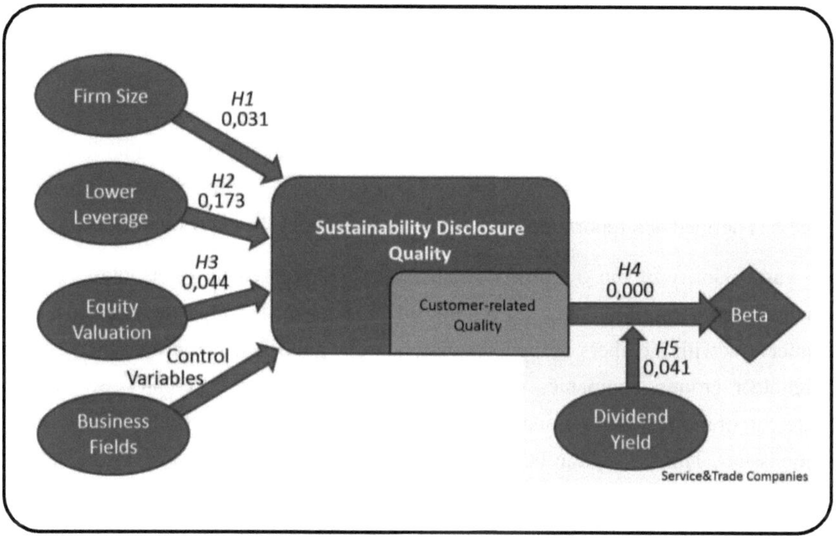

Figure 6: Determinants of Corporate Disclosure Quality and Consequences

4.4.3 Further Remarks on Corporate Disclosure Analyses

In addition to the quantitative results above, some interesting qualitative aspects related to the corporate reporting behavior of German mid-caps could be identified by conducting this study. It could be observed that nearly all corporations address employees in the introductive section of their disclosures. Usually, CEOs provide some basic information about their employees or at least thank them for their efforts in the editorial or foreword. In most cases, a separate chapter is devoted to employees.

With regard to diversity, companies that concentrate their business activities on the

German market tend to report less intensive although diversity is not limited to multi-cultural aspects. However, this subjective perception could not be measured comprehensively as most corporations of the sample are global players.

Corporations care about different stakeholders and their relevance for overall business activities. A few do even disclose materiality matrices to illustrate the relevance of specific issues to stakeholders compared to the relevance for the corporations. This allows deriving starting points in order to develop strategic plans. These matrices represent the corporate focuses well and are reflected in the corporate disclosures, too. One corporation even published an action matrix in its sustainability report to illustrate the corporate commitment to specific aspects evaluated by stakeholders and its influence on the corporate sustainability evaluation. By this method, the future decision-making process is defined and report addressees gain an idea about the corporate direction.

The vast majority of the examined disclosures are structured in a very classic way. However, one corporation does not provide a CEO foreword or investor letter. Instead, an interview with members of the managing board is published. Regarding the various stakeholder groups, companies seem to disclose information differently. Service & Trade corporations have a strong customer focus and report comprehensively on economic issues. This perception is also confirmed by the quantitative analysis as shown before.

However, corporations in general seem to disclose more information about sales activities in the business-to-consumer than in the business-to-business field. Maybe, the information costs of publishing insights on the business-to-business activities are too high as business-to-consumer activities can be observed well and the publication of corresponding issues is not connected to indirect costs. However, if this reason were true, the benefit of corporate disclosure would be limited significantly in this context. This aspect could be researched by further studies.

Most corporations disclose information about concepts to increase employees' motivation. Beside training and development opportunities that are mostly reported only to a vague extend, several corporations provide remuneration packages including share based items. However, one corporation even offers corporate bonds instead of a discount for its stocks. Based on the corporate financial situation, this cannot be seen as an innovative idea to increase employee participation but as a way to finance corporate activities. The singular assessment of this aspect by financial investors would be an interesting research question.

Several corporations publish information in compliance with the Carbon Disclosure Project. This project aims to disclose the total corporate emissions. As corporations seem to publish more numerical information on environmental issues in this context, many future-related aspects remain on a declarative stage. Overall, the fulfillment of current or prospective national and international regulatory stipulations seemed to be focused on.

In general, textile companies of the sample report more comprehensive on different social and environmental aspects with reference to their suppliers. In this context, economic parameters are not reported extensively probably to support the credibility of the stated information. Usually, economic aspects play an important role with reference to supplying parties and economic concepts are mentioned. This is even true for the field of nonphysical supply. The treatment of physicians in the medical area as business partners can be named as one example.

Especially the insurance corporations of the sample report in detail about market structures, trends and how they impact their business activities. On the one hand, this may have regulatory reasons as these corporations have to simulate different scenarios as part of the risk management and have to collect these data. On the other hand, these corporations usually have more information about their customers than other companies.

4.5 Discussion and Limitations

The aim of this first substudy was to investigate corporate sustainability disclosures of German mid-caps and to provide insights about determinants as well as potential consequences. To this end, a scoring model based on analytical grids was used. This scoring model was derived from previous research studies and theoretical concepts. It can be summarized that this approach works well and provides comprehensive results regarding the analyzed disclosures.

Firstly, determinants that drive the disclosure quality could be tested empirically and are conform to previous studies. Several interesting aspects with reference to the reporting intensity of specific issues could be elaborated. Secondly, financial consequences of corporate disclosure could be identified. Nevertheless, this section is not considered solely to provide an overview of the findings of this study. Moreover, the limitations of this study as well as potential research questions for future studies are discussed.

This substudy provides evidence that the quality of corporate disclosures of German mid-caps is driven significantly by the size and the market-to-book value of equity of corporations. The inverse impact of the leverage is non-significant and the consideration of the operating fields of the corporations improves the tested model significantly. It can be concluded that different determinants that represent corporate internal and external parameters should be used to explain the reporting activities.

The possession of necessary human resources to generate comprehensive disclosures is represented by the corporate size. This mostly internal parameter is marked by economies of scale effects. Compared to previous research approaches in the field of corporate reporting that focus primarily on larger corporations, size does also matter regarding the disclosure quality in the case of the researched German mid-caps. Beside internal economies of scale effects, the external expectation to larger corporations as well as their more relevant role for society and their corresponding higher exposition and public interest might explain this relationship.

In contrast to this, the market valuation of equity in relation to its book value can be seen as an external parameter that impacts the reporting activities. Based on the underlying concept of information asymmetries, the positive relationship between the market valuation of a company and its reporting quality is not surprising. Companies use corporate disclosures to signal quality and investors seem to reward this behavior trading the corporations with lower risk premia compared to others.

However, this link was not tested empirically as the market valuation of equity in relation to its book value was used an independent variable to explain the reporting quality. Nevertheless, the connection is likely to be circular.

As empirically tested, the market valuation drives the disclosure quality and companies are confronted with the concept of the information cost theory. They expect as one benefit from the publication of information to reduce their capital costs. With reference to corporate betas, the relationship could be tested significantly in the second part of this substudy.

The influence of the capital structure represented by the leverage is statistically not significant. Moreover, the operating fields that incorporate several different factors like the level of competiveness were used to improve the model testing of the disclosure quality determinants empirically.

All in all, the conducted test of determinants confirms the results of previous studies. This also supports the adequateness of the used scoring model as well as the (partial) transferability of gained knowledge from previous research activities to the mid-caps of the analyzed sample.

The detailed examination of the different parameters of the total disclosure score lead to some interesting insights that have to be mentioned. As the structure of the used analytical grids allows determining stakeholder- as well as dimension-related subscores, the corporate reporting behavior could be researched in detail. The analyzed corporations show two interesting reporting patterns. Firstly, the average disclosure quality with reference to supplier-related topics is lower compared to the other examined stakeholder groups. Secondly, economic issues are disclosed more detailed than social and environmental aspects.

Based on the theoretical foundation of corporate reporting, these deviations can be explained as follows. If corporations decide to publish information on a voluntary basis, they are confronted with direct and indirect costs. In the light of the information cost theory, they will only disclose information voluntarily if the foreseeable benefits outweigh the incurring costs.

Regarding supplier-related information, the direct cost that arises by the creation and collection as well as the verification and preparation of information might be higher compared to internal stakeholder groups. However, the force of this argument is limited as the customer-related reporting intensity is comparable to the employee-related intensity although customers also represent an external party. Consequently, this reporting behavior could be explained by focusing on high indirect costs, respectively a lower degree of benefits.

Indeed, the indirect costs are likely to be very high as particularly competitors can gain sensible information about strategic partners. Additionally, companies that have the intuition to offer future supply could adopt their contractual negotiation strategies with respect to the published information.

Simultaneously, potential benefits that may be generated by comprehensive disclosures with reference to supplier-related issues are relatively small. In general, companies expect to increase their reputation, lower their capital cost or avoid additional regulatory efforts by publishing voluntary information. Apart from the textile industry, reputational effects are in most cases of lower importance. Investors would care primary about suppliers if they acted monopoly-like or had a very strong negotiating power due to other reasons. Regulatory aspects are also of lower importance as suppliers are third parties.

It can be summarized that the lower level of the publication of supplier-related information can be explained by the concept of the information cost theory. While the arising indirect costs are comparatively high, potential advantages are restricted due to the role and classification of this stakeholder group.

In contrast to supplier-related information, the economic reporting dimension as measured by using the analytical grids, is comparatively high to other dimension-related subscores. Again, the information cost theory delivers the arguments to explain this reporting pattern. On the one hand, the direct costs to collect the relevant financial information are relatively low as most of the published economic information are easily available for the publishing companies and provided by the accounting departments. Indirect costs do exist although much information is already indirectly available due to the obligatory financial statements.

On the other hand, potential benefits can be regarded as comparatively high as financial investors are primarily interested in financial cash flows. Moreover, reputational effects might also play an important role because information that are related to financial issues can be regarded as costly and reliable signals and strongly support corporate strength. All in all, the high intensity of the published economic-related aspects in contrast to other dimension is not surprising but shows the rational consideration corporate entities are subject to.

In addition to the analysis of the determinants that drive the reporting quality and the elaboration of specific reporting patterns, the main contributions of this first substudy are the examinations of potential consequences due to the disclosed information.

While the information cost theory clearly states that benefits have to outweigh the direct and indirect cost of publication, the link between disclosures and corporate performance is not researched well and very unexplored with regard to mid-caps.

This study provides evidence that companies that operate in the service & trade industry can improve financial risk indicators by publishing comprehensive customer related information. To be more precise, it could be tested empirically that these corporations could reduce their unlevered beta factors that can be interpreted as risk indicators especially in downward markets. This also means that corporate reporting activities can be beneficial from a financial point of view.

Companies with lower beta factors are expected to have lower refinancing costs. This means that the risk assessment of financial investors is affected by the extent and quality of corporate disclosures.

As beta factors vary significantly over the different business fields, these tests were run on the different subgroups of the sample. Based on an extensive examination of previous research approaches, it can be concluded that this is the first study establishing the link between corporate sustainability disclosures and their impact on corporate beta factors of mid-caps.

Moreover, it could be tested that this effect is moderated by the level of the paid dividend yield. This means that companies can verify their published information and show their strength by sending this costly signal which is honored by a further decrease of the unlevered beta factors. Consequently, it can be summarized that the quality of corporate disclosures, subject to certain conditions, can impact the risk assessment of financial investors on the market level.

Nevertheless, several limitations regarding the sample, the research design and methodology have to be mentioned.

Firstly, the analyses were conducted on a sample that consists of 50 German mid-caps that are MDAX listed. The transferability of the findings to larger or smaller, non-listed companies and companies from outside Germany is questionable. Additionally, the research is limited to a specific point in time in order to analyze the consequences during a downward market trend of the MDAX.

Furthermore, the impact of the disclosures on the risk assessment could only be measured statistically for a sub-set of the analyzed companies and specific disclosure elements. In general, all tests are based on the corporate disclosure performance and not on an assessment of corporate activities.

Secondly, several technical and theoretical limitations have to be acknowledged. In order to determine the unlevered beta factors, aspects of the CAPM were used and therefore its critical appraisal has to be regarded from a conceptual point of view. Moreover, the additive inclusion of taxes is connected with additional assumptions.

As actual tax rates can vary significantly over time due to periodical tax loss compensation, average tax rates were used as input parameters to take the long-term influence of taxation into consideration. This should be relevant for corporations when they optimize their capital structure. Nevertheless, different parameters influence this process.

Furthermore, the impact of corporate disclosures on beta factors was analyzed by taking dividend yields into consideration. In fact, there could be additional parameters that might influence this relationship.

Nevertheless, several additional aspects are considered implicitly as stakeholder related scores were used to run the analyses for the different operating fields. Lastly, the used scoring model as well as each structural content analysis tool can be criticized. However, it is based on the methodological and theoretical basis of previous studies and research approaches and allows analyzing corporate disclosures along the value-added chain comprehensively.

Apart from the research results, this work also provides ideas and raises new questions for future studies. The following lines shall give an overview about potential tasks and research gaps that could be filled prospectively related to this substudy.

As most of the existing studies in this field focus on point in time analyses, there is a lack of panel data studies that examine reporting activities in the course of time. Especially with regard to the consequences of disclosures and its long-term effects, multiple year studies could add new knowledge. Additionally, the work on an international sample would also make potential consequences comparable as there might be country specific differences.

In contrast to long-term observations, studies that focus on corporate disclosures, their quality and consequences under very specific market conditions could also improve the understanding of corporate reporting and its assessment by investors. Situations like initial public offerings are characterized by a very high level of risk asymmetries and the influence of disclosures might be highly relevant.

Another interesting aspect could be the question whether the threat of additional regulation influences the reporting behavior or if financial crises have measurable effects and lead to long-term changes.

In addition to annual reports, companies publish an increasing number of specific disclosures like sustainability or CSR reports. Moreover, online reports as well as corporate websites are used to stay in contact with different stakeholders.

On the one hand, future studies could evaluate how reporting activities on the different channels are characterized and whether significant differences regarding their consequences do exist. On the other hand, the question of information overload may also become more topical. Furthermore, non-financial consequences or relations like the connection between customer satisfaction and corporate disclosures could be researched more detailed.

All in all, this substudy provides evidence that specific determinants drive the disclosure quality of German mid-caps and consequences with regard to the risk assessment of financial investors are measurable. Of course, the impact on the risk assessment could be researched more comprehensively as there exist many more potential risk indicators. These indicators could also be analyzed by further studies.

However, the focus on stocks has many advantages especially compared to corporate bonds. The relevance of ratings is less important, the influence of interest rate changes with regard to the concept of duration can be ignored and the capital structure can be incorporated quite well e.g. using the applied approach of this substudy. Just the market trend has to be taken into consideration to interpret the results of further studies that make use of beta factors.

5 Corporate Disclosures and the Decision-Making Process of Financial Investors

5.1 Framework

As the first substudy established a link between corporate disclosures and financial figures that represent the risk assessment of investors, a logical next step would be to research the decision processes of investors more comprehensively in order to understand the observed and tested phenomenon. Consequently, the aim of this substudy is to analyze the assessment of information from corporate disclosures and their impacts on the decision making of financial investors with respect to specific scenarios and personal attributes.

The overall concept of this second substudy is to capture behavioral and cognitive activities during the decision making process to invest in financial securities. In order to avoid the influence of other effects, specific scenarios that isolate realistic information that might be published through corporate disclosures were elaborated.

A core focus lies on the risk perception but also other aspects like ethical implications and their consequences to decision routines of private financial investors are examined. The results of this substudy might not only be helpful for companies to improve their reporting behavior. They might also support investors to objectify their decision processes and to overcome potential biases during their decision routines.

The composition of three main pillars sets the framework of this study. Firstly, the concept of risk and its inherency in decision making situations has to be regarded. In the context of this study the focus lies on financial risk and the exposure to lose money due to decisions and decision changes. As the measure of this kind of risk can be conducted objectively, the quantitative character of this work is supported.

Secondly, the moral philosophy type of the participants is taken into consideration. In order to follow the widely accepted classification of moral philosophies into teleology and deontology, the differentiation between the moral worth of behavior and methods to come to decisions is applied. The degree of moral influence into the decision process of financial investors may vary with the individual moral philosophy type. This type can be derived from the justification that is provided with the decision. As this philosophy type might also be relevant for the probability to change a decision if financial risk arises, a well-structured differentiation between teleology and deontology types is necessary. The used subcategories of teleology are egoism and utilitarianism. Utilitarianism can be further divided into act and rule utilitarianism. The corresponding categories for deontology are act and rule deontology.

Thirdly, the elaborated scenarios confront the participants with realistic dilemmas for financial investors that may arise due to new information provided through corporate disclosures. In order to provide a sufficiently large variety of situations, five different types of dilemmas are used. This variety allows researching the situational behavior of different moral types as well as the isolation of certain effects like the arising of financial risk and its impact with reference to the individual risk sensitivity and moral decision type.

The detailed theoretical background including the concept of risk as well the approach of different decision types are presented with the introduction of the tested hypotheses of this substudy in the following chapter. Afterwards, the research design section illustrates the methodology comprehensively and provides circumstantial information about the used scenarios.

In this context, the tested quantitative models of this study are introduced formally. This chapter is followed by the presentation of descriptive and inferential statistics as an overview of the research results of this substudy. Lastly, the conclusions are discussed and the potential theoretical and practical pertinence of this substudy is argued. Furthermore, the limitations of this approach are analyzed. The variable measurement in the context of this substudy is attached to this work (see appendix F).

5.2 Theoretical Development

In this chapter, the hypotheses which were tested empirically are formulated and derived from theory as well as previous research approaches. Moreover, the most relevant studies with reference to the hypotheses are briefly introduced. This section is divided into three parts. Firstly, the hypotheses to test the association between the decision and justification type as well as between the type and the probability to change the decision with reference to the different five scenarios are presented. Secondly, the determinants that drive the overall ethical decision making process are introduced. Thirdly, the hypotheses to establish the link between the intention to make up a moral decision and the willingness to change the decision due to negative financial consequences in the light of moderating effects referring to the degree of risk aversion are presented.

5.2.1 Theses Development of Scenario-based Behavior, Justification and Confidence

The decision behavior of the participants was tested over five vignettes that represent different dilemma types. As the individuals do not only have to agree with proposed decisions in the context of each scenario but also have to justify their choices among different response types, the elaboration of these response types had to be conducted comprehensively and is based on a strong theoretical background.

In order to develop the tested hypotheses that affect the association between the moral degree of the decision and the justification type as well as between the justification type and the probability to change a decision due to the threat of a pecuniary loss by the decision, a brief classification of the major philosophical theories that affect human behavior has to be provided. This summary shall support the process of operationalizing the normative theories to be able to collect relevant data and analyze them with quantitative methods.

Additionally, the concept of risk plays an important role in the context of this study. Consequently, the theoretical background of this aspect is explained and the practical relevance is illustrated with regard to a study published by Fraedrich and Ferrell (1992).

In order to structure the field of philosophical theories, many studies categorize them into teleology and deontology (Fraedrich and Ferrell, 1992). A classification on a further level allows deriving moral types that represent a sufficiently large choice among theoretically based decision types different individuals can identify with. Generally speaking, the main intention of teleology based theories is to capture the moral worth of decision behavior. The value that has to be maximized is the output of the behavior as this seems to be the only value-adding attribute and represents the total worth of behavior (Ferrell and Gresham, 1985). Teleology can be separated into egoism and utilitarianism. Utilitarian theories are further divided into act and rule utilitarianism.

While the goal of the egoist can be summarized briefly with the concentration on the own good and maximizing the own benefits ignoring any side effects, the utilitarian theories have to be regarded more detailed. Individuals that follow the act utilitarian approach tend to choose the activity that leads to the highest overall social outcome. In their eyes, outcomes that are inefficient for society can be regarded as suboptimal (Fraedrich and Ferrell, 1992).

In contrast to this, an individual that follows the rule utilitarian approach focuses on a potential rule that might be relevant for the activity and believes in reaching the best outcome through following this rule. This idea also contains a long time optimizing concept as a deviation from a short term optimum is accepted as following rules on the long run is believed to lead to highest benefits for society. Nevertheless, both types, act and rule utilitarians do not focus on their own benefits but try to maximize the good for society.

Deontology can also be separated into two large categories. Both do not follow the idea to maximize outcomes but to focus on methods and intentions that are related to the behavior in the context of the decision. Compared to the utilitarian theories, the focus moves from the societal to the individual level. The two categories are named act and rule deontology.

Act deontologists try to judge each situation individually as they believe this is essential due to its uniqueness. The consideration of the consequences is of less importance as the main focus is on a careful evaluation of situational factors. According to Fraedrich and Ferrell (1992) it can be summarized that for act deontologists these situational factors override potential rules and they asses situations based on a special moral intuition.

In contrast to this situational approach, rule deontologists regard the human capacity to evaluate each situation as to be too limited and consider the need of obligations that have to be fulfilled as essential. A widely known principle for instance is "Kant's categorical imperative" (Kant, 1972).

The idea behind this approach can be described by the approach to follow rules that everyone can agree with and is willing to follow. Rule deontologists regard this behavior as ethical and following these inter-cultural standards is interpreted as the best way to behave on an individual level. Fraedrich and Ferrell (1992) concentrate the rule deontologist view on the maxim that decisions are based on given rules and these rules override situations.

Based on this theoretical background and the results of previous studies like Fritzsche and Becker (1984) or Fraedrich and Ferrell (1992), rule-dominated individuals are expected to come to more ethical choices. This can also be formulated vice versa. Individuals, who consider moral aspects during financial decision process more comprehensively than others are expected to follow predefined ethical standards and should justify their decisions in accordance to the rule utilitarianism or rule deontology approach. Consequently, the following hypothesis in ascending order with respect to the hypotheses of the first substudy can be formulated:

H6: Individuals, who justify their financial decisions according to the rule utilitarianism or rule deontology approach, consider ethical aspects during decision processes based on corporate disclosures more comprehensively than others.

The hypothesis above will be tested separately for all five dilemma vignettes. With the inclusion of financial risk, another interesting parameter that might be supportive to understand the relationship between corporate disclosures and the individual assessment of provided information with reference to financial decisions is introduced.

As the definition of risk differs in the academic literature and with reference to the objectives of research studies, this term has to be defined precisely in the context of this study. While the first substudy focused on the empirical investigation whether a reduction of information asymmetries between investors and companies lead to a measurable market effect that incorporates the trading behavior of anonymous market participants following the understanding of systematic risk assessment (Sharpe, 1964), the second substudy also covers implicitly idiosyncratic risk characteristics.

In contrast to systematic risk, idiosyncratic risk can be eliminated through diversification and is not compensated pecuniary (Merton, 1973). Rational investors avoid taking any idiosyncratic risks. Nevertheless, it could be shown empirically, that idiosyncratic risk plays an important role in practice (Campbell et al., 2001) and its existence due to investor heterogeneity and further effects is proven (Goyal and Santa-Clara, 2003).

The approach of this substudy is to exposure financial investors to decision dilemmas in which they have to make up decisions that are associated directly with financial risk that is not compensated pecuniary. In contrast to the study of Fraedrich and Ferrell

(1992), this substudy is limited to financial risk. With reference to MacCrimmon et al. (1986), risk is defined as the uncertainty of suffering a financial loss in order to analyze the information assessment of investors with reference to provided information from corporate disclosures and their moral decision type.

In accordance with Fraedrich and Ferrell (1992), risk is inherent in the decision processes of human beings. They build up their work on the results of Rettig and Rawson (1963) who showed that risk is an relevant factor with reference to the engagement of unmoral behavior. Moreover, Zimmerman and Krauss (1971) elaborated significant associations between risk, the group membership and decision behavior with reference to unmoral choices.

Consequently, this substudy is based on the idea that financial risk is associated to decision behavior. As the concept of risk and decision types is tested on a sample of financial investors, the separation into groups is conducted on the provided response category with reference to each single scenario. With this procedure, the individuals who follow rule-based decision types can be observed for each vignette as they are expected of being less willing to change their decision with the introduction of financial risk as they tend to follow predefined regulations. Therefore, the second hypothesis of this substudy, which is again tested separately for the five vignettes, is the following:

H7: Individuals, who justify their financial decisions according to the rule utilitarianism or rule deontology approach, are less willing to change their decisions based on corporate disclosures with the introduction of pecuniary losses.

5.2.2 Theses Development of Determinants that impact Decision Process

The impact of personal attributes on the individual decision process has been tested in a series of previous studies even with reference to the applied theoretical model of dilemma vignettes by Husser et al. (2013). Due to the large amount of personal attributes that are tested in this substudy and the explorative nature of the research approach in the context of financial decisions and corporate disclosures, each hypothesis is introduced briefly providing an exemplary academic reference.

Previous studies researched intensively the relationship between moral aspects and the age of participants in heterogeneous frameworks. Peterson et al. (2001) focused on ethical beliefs of business professionals and discovered significant differences of morality between younger and older groups. As this finding might be transferable to investors with respect to this study, the following hypothesis is formulated:

H8: More life experienced individuals take moral aspects to a higher degree into consideration during the overall financial decision process than others.

Several empirical studies could find evidence that the return of value stocks outperform growth stocks in the long run and the risk that has to be accepted per unit of return of growth stocks is significantly higher (Fama and French, 2002). Growth investors are described to have the tendency of being short-term oriented and different motivations have to drive value and growth investors to make up their divergent investment decisions. With respect to the long-term perspective, value investors are expected to consider different investment aspects like ethical facets more comprehensively which leads to the following hypothesis:

H9: Value investors take ethical aspects into consideration when they come to financial decision based on information from corporate disclosures.

Eckbo and Verma (1994) showed empirically that the long-term development of the dividend yield is closely connected to the shareholder's voting right. Cronqvist and Nilsson (2005) proofed with their nested logit model that investors in general care about their influence on business activities of corporations.

Consequently, it can be assumed that those investors who emphasize the relevance of their right to codetermine the activities of a corporation and express not only their interest due to potential personal benefits but also their social obligation might take ethical aspects into consideration during their financial decision processes. Following this argumentation, this hypothesis can be derived:

H10: Investors who emphasis the relevance of their voting right are more likely to come to moral decisions in the light of dilemmas that are communicated through corporate disclosures.

Previous studies focused on a gender differentiation in the context of individual decision making. The clear majority elaborated differences that have been explained by genetic and socialization related aspects. Among many others, McMahon and Harvey (2007) showed empirically that female participants take moral aspects more comprehensively into consideration which leads to the following hypothesis:

H11: Female investors come to decision of higher morality than male investors when they are confronted with dilemmas.

Based on a sample of French buyers, Husser et al. (2013) researched attributes that drive individuals to come to ethical choices in the context of dilemma situations. They could show that the personal experience is one of two main factors that impacts the degree of morality that has been taken into consideration to make up the decision. They argue that experience makes individuals able to see beyond and withstand pressure. Following their argumentation and transferring their results to this substudy, the influence of experience can be integrated in the tested model through a hypothesis as follows:

H12: Investors with a higher level of trading and financial experience take moral aspects more into consideration than others when they evaluate decision dilemmas in the financial context.

The argumentation in favour of the formulation of the following hypothesis is very common to the one before and based on studies that link individual knowledge and experience with ethical decision making (Glover et al., 1997). Investors who reduce information asymmetries through the assessment of financial and economic news gain a better overview and understanding of critical situation and their impact on corporations. They know to differentiate between important aspects and noises that arise in the context of dilemma situations.

Consequently, these investors are expected to take ethical aspects into consideration to a higher degree and the following hypothesis can be formulated:

H13: Investors, who spend more time with the assessment of financial or economic news, take ethical aspects into consideration when they come to financial decision based on information from corporate disclosures.

The number of decisions that individuals have to make as well as their familiarity with decision dilemmas differs with reference to their professional environment. Gibson (2003) even describes that individuals adopt different moralities depending on the roles they fulfil. If participants transfer parts of their business role profile to financial decision situations, differences with reference to the evaluation of ethical aspects might be observable due to the professional status. The corresponding hypothesis is the following:

H14: The professional status is a relevant element of differentiation regarding the effect of ethical aspects on the financial decision process.

Hiltebeitel and Jones (1992) showed empirically in their study that ethical education contributes significantly to human decision procedures. The participants of their study relied on ethical principles like "the golden rule" when making moral decision after the integration of the corresponding educational content.

As it can be assumed that individuals who are well educated have been in contact with ethical theories directly and indirectly to a higher extent, this group is expected to rely on moral attributes also during financial decision processes more comprehensively. Therefore, the following hypothesis is proposed:

H15: Individuals with a higher educational level respect ethical aspects to a higher degree during the financial decision process than others.

5.2.3 Theses Development to link the Intention, Probability to change and personal Risk Aversion

The two hypotheses of this chapter link the a priori intention to consider moral aspects during decision processes with the exposure to financial risk. While the first hypothesis is limited to this interaction, the integration of the personal attitude to risk as a moderating effect is introduced in the second step. May and Pauli (2002) reviewed and investigated moral intentions and evaluations. They elaborated that moral intensity is related to the intention to consider morality during the evaluation process of the decision making.

Based on this approach, the participants of this substudy are expected to be less sensitive to the exposure of risk with reference to their willingness to change their decisions if they had the a priori intention to consider moral aspects during decision procedures. This can be justified with reference to the solid basis of their decision as well as with their personal conviction.

In contrast to this, participants with lower a priori moral intentions are assumed being more likely to change their decision with the introduction of potential negative financial consequences. The corresponding hypothesis is the following:

H16: Individuals, who have the intention to make up moral decisions on the basis of the information of corporate disclosures, are less willing to change their stated choice due to potential negative financial consequences.

The underlying concept of risk that is relevant in the context of this study has been described comprehensively with the derivation of hypothesis 7. It can be summarized that the focus of this study is on the quantitative financial aspect of risk and in accordance to Fraedrich and Ferrell (1992), risk is considered as being inherent in the decision making process of individuals if outcomes are not totally certain. Rettig and Rawson (1963) proofed risk as a relevant factor to engage individuals in different behaviours.

As the perception of risk may differ among persons due to divergent societal as well as personal backgrounds (Fraedrich and Ferrell, 1992) and the individual degree of risk aversion is an established concept to explain financial decision making with reference to functions that incorporate the utility respecting the uncertainty of outcomes (Diamond and Stiglitz, 1974), risk aversion is integrated in this substudy.

With reference to the reasoning of Diamond and Stiglitz (1974), participants that show higher levels of risk aversion are expected to try to avoid risks as they reduce their expected utility outcomes significantly.

In the context of this study, participants with higher levels of risk aversion are expected to be more likely to change their decisions although they had the a priori intention to make up moral choices with reference to the exposure to financial risk. This moderating effect of risk aversion is included in the following hypothesis:

H17: A decision change with reference to the a priori intention and financial risk is more likely if the individual shows a higher level of risk aversion.

5.3 Research Design

The methodological part of this substudy consists of three logical parts. Firstly, the detailed approach of the scenarios that have been used to for the conducted survey are explained. As this concept is derived from previous research work, the description is not only limited to the content of the different dilemma situation types.

Additionally, most important background information regarding these vignettes is provided. Secondly, the data collection and coding is described and further information regarding the sample that was the basis for the analyses is given. Thirdly, the used statistical methodologies are presented and the researched quantitative models are formally introduced.

5.3.1 Scenario-based Survey

The elaborated scenarios represent the core element of the survey. These scenarios are used as the most appropriate way to generate dilemmas which might occur in reality and financial investors could be confronted with. This methodological approach leads to the situation that the participants of the survey have to make up decisions based on isolated information which they have received from fictitious disclosures of corporations they can or already are invested in.

Consequently, comparative analyses can be conducted as the scenarios do not vary across the sample as well as isolated dilemmas are presented and the influence of secondary effects is eliminated (O'Fallon and Butterfield, 2005).

The scenarios are based on the work of Fritzsche and Becker (1984) and follow their five dilemma vignettes. These are in alphabetical order:

- Coercion and Control
- Conflict of Interest
- Physical Environment
- Paternalism
- Personal Integrity

These five dilemmas are constructed to represent a large variety of decision problems. They have also been used for other academic studies (e.g. Husser et al., 2013) and are consistent with other theoretical concepts to observe moral maturation and conation (Hannah et al., 2011). For each scenario, the participants are asked to which degree they agree with a described decision.

In combination with the classification of the responses of the participants due to the before mentioned moral philosophy types, the decision can be analyzed with regard to types and personal attributes. Through the collection of further data, the probability to change a decision that has been made can be analyzed with reference to the decision type.

As the primary focus of this chapter is on the detailed explanation of the applied scenarios, the following lines contain important information regarding the background as well as the concrete design of the dilemma vignettes. The theoretical background of the response categories that are directly connected to the moral decision types are explained comprehensively in the corresponding theses developing sections.

The coercion and control category represents a dilemma that forces the participant by external factors to make decision by using threats, extortions or comparable uses of power (Fritzsche and Becker, 1984). In the context of this study, the decision maker is confronted with an annual report that contains the information about a potential financial aid for a developing country.

Additionally, sensitive information based on a confidential letter from the CEO of the company is given. The participant is informed that the market entry to a developing country with huge potential is connected with a payment in bribes. Without this payment the local authorities will not authorize any business activities in this country. However, this payment which will be declared as a developing assistance will only be made if the majority of the investors agree with it.

Consequently, the decision maker is confronted with a dilemma that might lead to a potential personal benefit through the acceptance of an unmoral behavior.

The decision that has been made can be justified through the selection of one out of five response categories. Each of them represents a decision type. In the case of the coercion and control category, the rule utilitarian type does not agree with the payment as there could not be any trust if it were common to pay in bribes.

In contrast to that, the act utilitarian agrees with the payment with the argument that it were beneficial to make the payment in order to increase the competition as well as the overall circumstances in the developing country. The rule deontologist ignores potential positive benefits of paying in bribes and argues that this would not be correct from a moral perspective.

Lastly, the egoist just cares about his own benefits and follows the argumentation that the decision to make the payment leads to positive benefits for the company. In order to proof the solidness of the made decision, the coercion and control scenarios like each vignette closes with the question how likely it would be to change the decision if the participant may be exposed to a pecuniary loss.

The conflict of interest category represents the second dilemma vignette. The participants are confronted with situations in which they have several interests that are of contrary nature. The decision in favor of their own good might be disadvantageous for other parties that are involved directly or indirectly.

The concrete dilemma of this study describes a foreword that is part of an annual disclosure. In this text, the chairman of the supervisory board informs the investors that the corporation plans to construct a new plant and positive as well as negative aspects are illustrated. This project would lead to a substantial increase in the market share and positive tax benefits can be realized. Due to a higher level of automation, the number of employees could be reduced which would lead to positive effects on the profit margin.

Nevertheless, the realization of the new plant demands the investment of considerable resources. The participants are asked to state to which degree they agree with the investment in the new plant in the light of the advantages and disadvantages.

Again, the given answer can be justified with five different response categories. The order of the decision types is changed to avoid the reselection according to the same pattern. The egoist argues that the decision to make the investment would lead to an increase of the profitability of the corporations. As this generates additional value for the investors, the egoist follows the argumentation to invest substantial resources in order to build the new plant. Negative aspects are not mentioned.

In contrast to this, the rule deontologist refuses the investment although potential benefits are observed. The reasoning is not based on arguments but follows a principle based practice. The act utilitarian does not follow such a principle but focuses on the output of his decision. In this case, the act utilitarian argues that the investment would be beneficial as it increases the overall competition. This justification is not limited to the investor himself but also focuses on potential benefits for other parties.

In contrast to this, the act deontologist does not see any obligation and need to make the payment and rejects the project. The argumentation of the rule utilitarian is much more driven by ideological ideas. The investment is denied with the reason that the focus on the market share cannot be regarded as a dominating principle as this would not lead to higher happiness for the society. In analogy to the other scenarios, the willingness to change the decision due to a potential financial loss that might occur is evaluated.

The physical environment dilemma type represents the third vignette. In accordance with Fritzsche and Becker (1984) this is a special conflict of interest dilemma that affects the natural environment.

In most studies, negative consequences for the nature have to be balanced against potential positive economic benefits (Husser et al., 2013). This study describes information that is published in the sustainability report of the corporation the participants are said to be invested in. This disclosure contains data about the emissions due to the production process of the corporation. It can be observed that the emissions grow steadily but are still below the legal limits.

Moreover, the investors are informed that the emissions could already be reduced through the introduction of eco-sensitive production processes. However, this introduction would lead to more cost-intensive production processes. The participants are asked to which degree they agree with the introduction of environmental friendly production processes that reduces the profit margin of the corporation.

The participants can follow the argumentation of the rule deontologist who is willing to accept the investment in environmental friendly production processes. Although he considers negative consequences to the profit margin, the rule deontologist is willing to accept these costs in favour of a moral decision.

The approach of the act deontologist varies from this position. As there is no legal obligation to make the investment in environmental friendly production processes yet, the act deontologist rejects the investment offer. The argumentation of the act utilitarian differs from this position. In order express a given level of uncertainty, he argues that in some situations it might be more beneficial to focus on the essentials as this leads to a greater disposable income. In awareness of the fact that a disposable income could be attributed flexibly and raises the benefit of a society, the act utilitarian rejects the investment.

The position of the egoist is less complex and can be summarized easily. In his eyes, the investment reduces the competiveness of the corporations he is invested in. In the light of a lower profit margin and potential other negative financial effects, he is not willing to agree with the introduction of more cost-intensive production processes to protect the environment.

The reasoning of the rule utilitarian to accept this change follows the intention that through the focus on the improvement of the working and living conditions, a better society can be generated. In analogy to the other dilemmas, the final scenario related question discovers the willingness of each participant to change his decision due to a potential pecuniary loss.

The paternalism vignette deals with an issue that affects the welfare of external parties and autonomous decisions of the group the participant is assigned to. The concretion of these decisions impacts directly the situation of external parties not without consequences for the group's status.

In the context of this study, the participants are confronted with an important chapter of the current annual report of their corporation. The problem of a defective lot of products is described in this section. As these products were delivered to customers, no more defective products are on stock of the corporation. There is no information about the defect and its potential consequences provided.

However, the participants can read that the customers are not able to make any legal claims against the corporation due to limitation periods that have been reached. Consequently, the corporations cannot be forced to repair or replace the defective lot of products. The participants are asked to which degree they agree with the proposal of voluntary compensation payments in favour of the affected customers.

The act deontologist rejects the proposal of voluntary compensation payments and focuses on the legal situation. As long as there is no legal obligation to make the payment, he would not be willing to accept benefits in favour of external customers on the company's account.

The rule deontologist does not ignore the monetary consequences to the corporation but agrees with the idea of voluntary compensation payment with the argument that this payment is correct from a moral point of view. As the egoist is primarily focused on his own welfare, he does not share this attitude. Participants who follow the argumentation of the egoist reject the payment as it is costly and might also reduce the dividend yield.

The approach of the rule utilitarian is divergent. He does not focus on the negative aspect for the corporation but prefers to make the compensation payment in order to have peace of mind. Participants who follow the rule utilitarian statement interpret the fairness to compensate an own mistake as an unwritten rule that must not be contravened.

The act utilitarian is less focused on the fulfilment of rules but interested in the concrete output of the actions. Consequently, participants that follow his points refuse the voluntary compensation payment to customers as it seems to be more beneficial to invest the money for more useful purposes. Concrete other purposes are not illustrated but might be internal or external with reference to the corporation.

Again, this vignette closes with the question if the participants would be willing to change their decision in the light of a potential pecuniary loss.

The last dilemma type is called personal integrity and affects the conscience of the participants. This vignette generates the situation that the participants have to conduct a self-evaluation regarding their own values and objectives in order to come to a decision that affect their own welfare as well the welfare of the society. The transparency of the own acting is not described in detail but can be considered as being on a low level.

In the concrete scenario of this study, the corporation publishes a social responsibility report on a regular basis. The individual participant is said to be able to identify that most of the disclosed information deviate from real conditions. Further information about negative consequences to employees or other stakeholders are not described. The source of information of the participant concerning the real conditions is not named.

Additionally, the scenario describes that the participant has already be in contact with a leading member of the reporting unit after reading the disclosure. This phone call leads to the result that the reporting unit is aware of the fact that incorrect circumstances are reported but prefers to communicate a positive image to its external stakeholders. As the participant knows a reliable investigative journalist since his school years, the question arises to which degree he agrees with the idea to contact him in order to make him aware of this fact.

The rule utilitarian agrees with the proposal to inform the journalist about this incongruity. Participants who follow his argumentation agree with the general idea that there could not be any trust, if it were common to cheat on reports.

This reasoning reflects the ideology of the rule utilitarian very well. The act utilitarian makes a different decision and rejects the proposal to inform the journalist. He argues that it is beneficial to communicate good circumstances and to improve the image of the corporation as this leads to a better access to financial resources. The positive consequences of the reporting behaviour override a principle driven evaluation of the situation.

The rule deontologist refuses this behaviour and speaks in favour of informing the journalist as cheating on reports is not correct regardless any potential positive benefits that may arise. The act deontologist follows a divergent argumentation. In accordance to his approach, there is no legal obligation to report extensively and correctly about sustainable issues and the incongruity between the disclosed and real conditions is accepted due to its positive benefits for the corporation. Moreover, the incorrect disclosures do not harm anyone at first sight.

The egoist agrees with the idea to generate a positive image of the corporation and is not willing to inform the journalist. However, his decision is driven by the reasoning that a positive corporate image leads to advantages for the corporation which will be beneficial for the shareholder. The vignette closes as usual with the evaluation if the participant would be willing to change his decision due to a potential monetary loss.

5.3.2 Data Collection

In addition to the above described scenarios, several information regarding personal characteristics as well as behavioral self-assessments were collected in the context of this substudy. The data collection was organized the following way. Firstly, the participants were asked to provide several personal attributes. Due to an additional investigative question, the degree of personal uncertainty avoidance was tested. Secondly, the above described scenarios were introduced. Finally, the relevance of risk and moral attitudes were researched.

The collection of personal attributes was not only limited to age and gender related indications. The educational level was inquired with reference to four given ascending categories.

Additionally, information about the professional status again divided into four groups of categories was collected. The self-assessment of the financial/trading experience, the personal relevance of the voting right as well as the personal relevance of moral issues in the investment process were recorded on two 10 point Likert scales.

Moreover, the daily time of the consumption of financial and economic news was asked. The differentiation between investor types (growth or value investor) was conducted using an investigative question about investment selection criteria. The last item of the first question section was initiated to illuminate the degree of risk aversion through offering to enter a risky investment opportunity with a selectable initial investment amount.

As the scenarios and the corresponding questions were described in detail in the previous chapter, the information regarding the data collection can be named briefly. The agreement with the proposed behaviour as well the willingness to change a decision with the introduction of financial risk were recorded on 5 point Likert scales.

Additionally, the decision types are assigned to the selection of the provided justification categories. Finally, information about the degree to avoid any sorts of risks as well as the general relevance of morality for decision processes were collected using two 5 point Likert scales.

The concept and articulateness of the survey was analyzed through the application of several pre-tests. As the survey was carried out during a fair, the participants were not asked for self-written answers due to practical reasons as well as to ensure the anonymity of the participation.

The researched sample consists of individuals who attended the "Boersentag" in Hamburg which is one of the largest trades for financial investors in Germany. It took place in late October 2014 and attracted around 5500 visitors. The original survey text was in German.

5.3.3 Data Coding

The methodological approach of this substudy is shaped by the idea of deriving new information through the application of statistical procedures on the collected data. In order to prepare the collected information so that they can be analyzed statistically, several transformations are necessary to facilitate data analytical tests. As this categorization has to follow clear guidelines, the process of data coding is described in the following lines.

Firstly, a consecutive numbering of the participants is initiated in order to guarantee the correct assignment of personal attributes as well as decision related information to each individual. Apart from the age, the income and the economic/financial information time that are taken without any transformation, a data coding is necessary to capture the other information. The gender is binary coded; the educational level as well as the professional status are divided into four categories.

In order to capture the trading/financial experience, the voting relevance as well as the importance of morality 10 point Likert scales are used. The investor type and the degree of risk aversion are divided into three groups. The personal relevance of risk and the perception of the importance of morality in the context of decision making are measured on 5 point Likert scales.

For each scenario, the level of the moral degree is recorded on 5 point Likert scales. The total moral decision level over all five scenarios is binary coded based on the subresults of each vignette and their aggregation. Additionally, numerical attributes are assigned to the decision types that are divided into five different groups and also gathered in this context.

5.3.4 Sample Description

As the aim of this substudy is to derive universally valid results among different investors groups, the selection of the participants was driven by the idea to compose a representative sample. Consequently, a sufficiently high heterogeneity among obvious criteria like the age and gender was aimed during the selection process. Finally, 146 individuals participated in the study.

Out of these 146 participants, 64 (43,84 %) are female and 82 (56,16 %) are male. The average age is 47,38 years; the median is 42 years. With a skewness of 0,547 (standard error of skewness: 0,201) the data can be used well from a data analytical point of view. With a range from 18 to 87 years, the age coverage over the adult population is excellent.

Although the study was conducted anonymously, the participants have been very restrictive regarding information about their income and did not submit information in all cases. Consequently, this variable is not used for statistical analyses as the precision of the estimation of missing values cannot be guaranteed sufficiently to conduct reliable multivariate analyses.

Over all five scenarios, 70 (47,95 %) participants make up predominant ethical choices and 76 (52,05 %) individuals do not. As the determinants of this decision behaviour are analyzed in detail in the corresponding section of this substudy, no more information are provided in this context.

Additionally, scenario specific information with reference to the researched sample is provided in the context of the presentation of the results of the scenario-based hypotheses.

Based on the conducted classification of the investor types regarding the participants of the sample, around 44 % are value investors, around 46 % are growth investors and around 10 % could not be classified definitely. While 52 % of the female investors indicate being value investors, only 38 % of the male investors follow this approach. In contrast to this, 31 % of the female participants agree with the growing investor approach and 57 % of the male investors do so. 17 % of the female investors cannot be classified clearly and only 5 % of the male investors are not grouped. As there seem to be gender related differences, further analyses consider this parameter more precisely.

On average, the researched participants spend around 13,85 minutes per day on consuming financial or economic news. Surprisingly, the group of persons with the lowest educational level is the group with the highest average time and spend 21,38 minutes daily.

However, this group is the oldest group with an average age of 61,46 years with reference to the educational level as the grouping parameter. In combination with the professional status, it can be assumed that most individuals of this group are retired and able to invest an above average time on consuming financial or economic news.

Nevertheless, it can be summarized that in this sample there is no link between a higher educational level and the time invested gathering financial news.

While the provided trading experience seems to be very symmetrical with an average of 5,05 on a 10 point Likert scale and a skewness of 0,007 (standard error of skewness: 0,201), the relevance of the voting right is of high importance to most individuals of the sample.

In order to measure this parameter, a 10 point Likert scale is used, too. However, the average relevance is 7,10 with a skewness of -1,146 (standard error of skewness: 0,201) and indicates in combination with a comparable variance the outstanding meaning of this parameter. The voting right seems to be slightly more important for female investors with an average of 7,38 than for male investors with an average of 6,89.

Surprisingly, the relevance of the influencing power of the voting right on the business activities of the corporation does not differ significantly between value and growth investors. However, participants with an above average age declare a marginal higher relevance of the voting right (7,35) than the group of younger individuals (6,92).

The degree of risk aversion is measured on a three level scale. The group with the lowest risk aversion consists of individuals with the lowest average age (41,28). The average ages of the group with the moderate risk aversion (49,47) and the high risk aversion (49,74) do not differ significantly.

Further analyses with reference to the intention to make up moral choices and the willingness to change decisions with the introduction of financial risks are provided in the corresponding section of the hypotheses testing.

5.3.5 Statistical Analyses and Models

The different hypotheses of this substudy are tested through the application of multivariate statistical procedures. Due to formal requirements and to provide a comprehensive understanding of the conducted data analytical tests, the corresponding models are presented in this section. In analogy to the first substudy of this work, widely known statistical methods are described very briefly as the focus is on the concrete design of the tested models.

The structural setting of this chapter follows the sequential arrangement of the hypotheses testing. It starts with the scenario related analyses and models and continues with the overall decision related ones. Finally, the intention, changing probability and risk related modeling is presented.

For each of the five scenarios a predefined test sequences is stipulated that guarantees the statistical correctness of the applied analyzing methods as well as the delivery of reliable results.

In order to proof the association between the ethical degree of the made decision and the decision type as well as the association between the decision type and the likelihood of a decision change, the following test procedures are run. First of all, the homogeneity of the variances is tested using the statistic of Levene. As this test statistic is well known, its formula is not presented in the context of this study.

With the application of this inferential statistic the hypothesis is tested that the variances are equal which indicates the need of the application of generalized test methodologies. In the context of this substudy, the applied significance level is 0,05 which is used as a termination criterion. As significance levels below 0,10 are also considered to be very critically, a rechecking analysis was planned. However, in the context of this substudy there is no case that needs the execution of this rechecking procedure.

As a second step, the standard procedure of an analysis of variance is run in order to test potential differences between the means of the observed groups. As the applied method is a standard way of additive data decomposition followed by the comparison of mean squares using F-tests, the correspondent statistical background is not described in detail.

It can be summarized that the F-test compares the factors of the total variance and delivers reliable results respecting the high level of degrees of freedom (141) in the context of this substudy. The aim of this approach is to test the variances of the moral degree with reference to the decision type and with another separated step the variances of the decision type and probability to change the decision. In order to derive results with a sufficiently high correctness, the level of significance is fixed on 0,05.

For those cases that reject the Levene statistic and show significant results for the analysis of variance, two different kinds of post-hoc tests are run in order to conduct a group separating with reference to the divergent group means.

On the one hand, Duncan's new multiple range test is used as a very sensible statistical procedure to separate into groups. On the other hand Tukey's honest significance test as an approach to prevent false positive errors is executed. Consequently, the results of the testing can be interpreted in the light of the tendencies of the tests to generate type I and type II errors.

The Duncan test consists of a series of pairwise comparisons of the means. The comparison is conducted on the level of significance of 0,05 in the context of this study and is performed sequentially following a predefined order confining the data set. Differences are significant if they exceed the mathematical product of the significant studentized range and the standard error.

In contrast to this, the applied Tukey test determines a value, the so called honest significant difference, which represents the distance between the groups which has to be exceeded so that group means are regarded to be sufficiently different.

Lastly, correlations are tested in the context of the different scenarios in order to proof decisions with reference to personal attributes. Regarding the divergent scale levels, specific rank correlation coefficients have to be calculated.

To this end, Kendall tau rank correlation coefficient as well as Spearman's rho are used in this context as they are nonparametric measures to proof statistical dependence between variables. Spearman's rho follows the basic idea of this approach to evaluate the fitness of a monotonic function in order to describe the relationship between the tested variables.

The selected significant level to judge the correctness of the calculated correlation is set on 0,05. The level of significance is equal in the case of Kendall's tau. However, the methodological approach differs: Kendall' tau compares the observations pairwise and they are declared as concordant if the ranks of both of their elements agree. Vice versa, they are noted to be discordant.

The number of concordant and the number of the discordant pairs as well as the total number of cases are the relevant input factors to calculate the corresponding correlation coefficient. As Kendall's tau converge a normal distribution more efficiently than Spearman's rho, both correlation coefficients are calculated and interpreted in the context of this substudy.

The overall decision behaviour that can be observed with reference to the five vignettes is researched using two methodological approaches to calculate discriminant functions as well as through the application of a binary logistic regression in order to test parameters that drive ethical choices.

The methodological difference in order to calculate the discriminant function refers to the way to conduct the separation. Firstly, the Mahalanobis distance (m) is used; secondly Wilk's lambda (w) is applied. The coefficients (b) and (a) are calculated to estimate values of the discriminant function that separates the groups clearly.

As both methodological approaches lead to congruent results that are presented comprehensively in the corresponding section, the discriminant function form coincides with reference to the introduced methodology-related indices and is written formally as follows:

$$d_{m;w} = b_1x_1 + b_2x_2 + b_3x_3 + b_4x_4 + a. \quad (8)$$

In the case of the Wilk's lambda approach, the a priori probabilities of all groups are set to be equal in the beginning. The covariance matrix is used within the groups. Broadly speaking, the test statistic of Wilk's lambda is applied to evaluate whether average values of the discriminant function differ significantly. Moreover, analyses of variances are calculated to validate the results.

The canonical correlation as a reliable method to evaluate cross covariance matrices is determined in order to estimate the quality of the calculated function. This value represents the correlation coefficient between the values of the discriminant function and the decision group membership.

Canonical correlations above 0,5 are regarded as sufficiently high in the context of this substudy. Additionally, the Eigenvalue is taken into consideration to judge the separation quality of the discriminant function. To account for outliers, the squared Mahalanobis distances to the centroid that represents the mean value of the discriminant values are calculated.

The second methodological approach to determine the discriminant function uses the Mahalanobis distances as the grouping criteria. In this context, the execution of the Mahalanobis distances stepwise method leads to a maximization of the distance between the two groups.

In analogy to the Wilk's lambda approach, several factors are observed to control for the grouping quality and significance. In both cases, not only the coefficients but also the standardized values are shown and the relevant histogram that illustrates the distribution of the discriminant function is presented together with the results in the hypotheses testing section.

In the case of the binary logistic regression a stepwise method is applied following the forward selection approach. This means that the initial model starts without any variable and tests the addition of new variables. To be more precise, the correlation of variables to the independent variable is taken as a selection criterion and with the application of the Wald test, the contribution of the variables is analyzed in order to remove those that do not support the explanatory power of the tested model.

A linear predictor function for each data point can be derived that consists of the regression coefficients that indicate the relative effect of a personal attribute on the overall made decision.

As the coefficients are equal to those that are tested in the context of the discriminant function, the linear predictor function is formally introduced in its simplest form using the notation of a scalar product of the vector of the regression coefficients and the vector of the explanatory variables. The formula of the linear predictor function for each data point (dp) including the regression coefficient vector (β) and explanatory variable vector (x) is:

$$f(dp) = \beta_3 \times x_{dp}. \tag{9}$$

The effect of the intention to make up ethical choices during the financial decision process on the changing probability in the light of the introduction of financial risk is tested in the context of a linear regression set.

In order to include the individual degree of risk aversion, a moderator analysis using a trichotomous moderator is conducted. To make the influence of the moderating variable measurable, a two block set up is created to analyze the explanatory power of the model including and excluding risk aversion.

With reference to the different levels of scale, standardization procedures are run. The moderator is defined as the mathematical product of the standardized intention Zscore(int) and the standardized degree of risk aversion Zscore(ra). The abbreviation of the arithmetic mean of the willingness to change the decision is (CAM). Formally, the following equation describes the conducted moderator test:

$$C_{AM} = \beta_{40} + \beta_{41} Zscore(int) + \beta_{42} Zscore(ra) + \beta_{43}(Zscore(int) \times Zscore(ra)) + \varepsilon_{4t}. \quad (10)$$

5.4 Research Findings

The research findings of the second substudy are presented in this chapter. The structure follows the numerical order of the hypotheses. In the first subsection, the results referring to the scenario related hypotheses are presented including pre-tests to analyze the data structures in order to ensure the application of adequate statistical methods. The second and third section describes the results of more sophisticated approaches related to the overall relationship between corporate disclosures, personal characteristics and financial decisions.

These chapters also provide information about the adequateness of the underlying econometrical models and the confirmation or rejection of the underlying hypotheses of this substudy.

5.4.1 Testing of Scenario-based Hypotheses

In this section, the quantitative tests regarding the individual decisions and the provided justifications as well as the probability to change the made decisions referring to the decision type with the introduction of financial risk are described. These results are related to hypotheses 6 and 7 of this work. The hypotheses are tested separately for all five dilemma vignettes.

The structural procedure is the following: Firstly, the association between the degrees of moral decision and the justification type are tested. Secondly, the link between this decision type and the probability to change the decision with the introduction of financial risk is proven. In both cases, tests to check the homogeneity of the variances are conducted. Post-hoc tests are used to create groups with reference to the decision behavior. These tests are always conducted on a 0,05 level of significance.

Additionally, correlation analyses are run in order to check for associations between personal attributes and decision patterns.

The results of the conflict of interest scenario are clear without any doubt. The Levene statistic (0,853) regarding the ethical degree of the decision is not significant (0,494). Consequently, the analysis of the variance can be used for interpretative purposes. The moral degree differs significantly ($F = 28,048; 0,000$).

Tukey HSD divides the act utilitarian and the egoists in one group (0,999) with a significant lower ethical level and the other decision types into another (0,873). This result coincides with the analyses according to Duncan. The act utilitarian and the egoist show a lower level and build an own group (0,813) and the other justification types are separated (0,371).

Kendall's tau correlation (0,274) as well as Spearman's rho (0,353) indicate a significant positive relationship between the moral degree and the age. The Levene statistic (1,888) for the tests between the decision type and the willingness to change the decision with the introduction of financial risk is also not significant (0,116). Again, the analysis of the variance is significant ($F = 22,505; 0,000$).

Tukey HSD shows that the rule deontologists and the act deontologists (0,965) are less willing to change their decision with the introduction of financial risk than others (0,944). The calculation of the Duncan statistic confirms this separation (0,512; 0,484).

The results of the physical environment scenario show some differences. With reference to the ethical degree and the justification type, the Levene statistic (1,019) is not significant (0,400).

The analysis of variance shows a significant (0,012) result (F = 3,53). However, the results of the post-hoc tests differ. According to Tukey HSD, act deontologists, act utilitarian and egoists build a group with a lower level of morality (0,205) compared to the other justification types (0,188).

The separation according to Duncan creates even three groups whereby the mentioned types build up the lower two groups (0,410; 0,188) in contrast to the rule deontologists and rule utilitarian (0,416). Significant correlations could not be identified in the context of this vignette.

With reference to the association between the decision type and the willingness to change the decision, the Levene statistic (1,552) is not significant (0,191). However, the analysis of the variance does also not deliver a significant result (F = 0,921; 0,454). In this case of environmental issues, no significant differences regarding the willingness to change the decision can be observed with reference to divergent decision types.

Thirdly, the results of the paternalism vignette are presented. For this scenario, the Levene statistic (1,018) is not significant (0,400) with reference to the decision level. The analysis of variance referring the moral degree of the decision is also not significant (F = 0,549; 0,700).

Consequently, the post-hoc tests do not provide any useful information. Correlations between personal attributes and decision related factors could also not be identified. The test of homogeneity of the variances regarding the decision type shows a non-significant (0,670) Levene statistic (0,590). The analysis of the variance is significant (F = 4,042; 0,004). Both post-hoc tests deliver consistent results.

Tukey HSD divides rule deontologists and act deontologists in one group of lower willingness to make a decision change (0,077) compared to the other decision types (0,966). The calculation of the Duncan statistic confirms this separation (0,970; 0,543). Further significant correlation between personal attributes and decision patterns could not be identified.

The Levene statistic (0,732) with reference to the ethical choices of the personal integrity scenario is not significant (0,572). As the analysis of variance does not provide any useful result (F = 0,875; 0,481), the post-hoc tests are not useable.

With reference to the test procedure of the decision type and willingness to change the decision with the exposure to financial risk, the test of the homogeneity of the variances is even significant. Consequently, further analyses are not run in order to avoid potential misinterpretation based on biased statistical tests. All in all, the results of the personal integrity scenario are not satisfying.

In contrast to the personal integrity scenario, the coercion and control vignette provides several interesting and significant insights. The Levene statistic (2,152) is slightly not significant (0,078). The results of the analysis of variance of the ethical decision degree shows a highly significant result (F = 28,393; 0,000).

The results of both post-hoc tests are consistent. Tukey HSD divides act utilitarian and egoists in one group (0,951) that represent a lower level of moral integrity regarding the assessment of the scenario compared to the other justification types (0,931). The calculation of the Duncan statistic reassures this grouping (0,472; 0,458). Both rank correlation coefficients, Kendall's tau correlation (0,320) as well as Spearman's rho (0,425) indicate a significant positive relationship between the ethical degree and the age.

The Levene statistic (0,878) with reference to the willingness to change the decision is clearly not significant (0,479). The analysis of the variance provide a significant result (F = 27,202; 0,000).

Again, the separation according to Tukey HSD and Duncan are consistent. According to Tukey HSD, rule deontologists and act deontologists create a common group (1,000) that is less willing to change the made decision compared to the other decision types (0,975). This result is confirmed by the calculation following the approach of Duncan (0,972; 0,575).

It can be summarized over all five scenarios that participants who justify their financial decisions according to the rule utilitarianism or rule deontology approach consider ethical aspects during decision processes based on corporate disclosures more comprehensively than others.

However, it has to be mentioned that act deontologists show significant lower morality only in one scenario. Nevertheless, hypothesis 6 has been confirmed. Additionally, it could be shown empirically for two independent scenarios that act deontologists and rule deontologists are less willing to change a made decision than other decision types.

Consequently, the proposed hypotheses 7 can only be confirmed partly as individuals who follow the rule utilitarianism approach could not be tested to be less willing to change their decision with the exposure to financial risk. Consequently, the first part of the hypotheses 7.1 that refers to rule utilitarianisms has to be rejected; the second part of the hypotheses 7.2 that refers to rule deontologists can be confirmed.

5.4.2 Testing of overall Decision-Making Hypotheses

The impact of several personal attributes of the participants has been tested on the overall moral decision behavior. The results of the discriminant analyses with reference to the two different methodological calculation approaches as well as of the binary logistic regression are presented in this chapter.

In all cases, the binary coded dependent variable is 0 for unethical and 1 for ethical choices. Firstly, the discriminant function is calculated using the Mahalanobis distances as the grouping criteria. The pooled within-groups matrices of the independent variables are identical as the analyses were run on the same data. They are listed in the following table:

Table 6: Pooled Within-Groups Matrices

	Age	**Inv-type**	**Vo-ting-rel**	**Gen-der**	**Tra-ding-exp**	**Info-time**	**Prof-sta-tus**	**Educ**
Age	1,000	-,044	,032	,064	-,057	,254	,476	-,112
Inv-type	-,044	1,000	,085	-,001	,203	,031	-,018	-,015
Vo-tingrel	,032	,085	1,000	-,024	,184	,040	,046	-,062
Gender	,064	-,001	-,024	1,000	-,115	-,228	-,066	-,028
Tra-ding-exp	-,057	,203	,184	-,115	1,000	,145	,082	-,024
Info-time	,254	,031	,040	-,228	,145	1,000	,232	-,104
Prof-status	,476	-,018	,046	-,066	,082	,232	1,000	-,222
Educ	-,112	-,015	-,062	-,028	-,024	-,104	-,222	1,000

As described in the statistical methods and models section, the applied method follows a stepwise procedure. The following table provides information about the single steps, the correspondent statistics and the entered independent variables:

Table 7: Variables entered/removed through stepwise Procedure

Step	**Entered**	**Min. D Squared**					
		Statistic	Between Groups	Exact F			
				Statistic	df1	df2	Sig.
1	Infotime	,640	uneth. and eth.	23,327	1	144,000	3,455E-006
2	Gender	1,025	uneth. and eth.	18,548	2	143,000	6,889E-008
3	Votingrel	1,337	uneth. and eth.	16,019	3	142,000	5,035E-009
4	Age	1,597	uneth. and eth.	14,245	4	141,000	8,637E-010

At each step, the variable that maximizes the Mahalanobis distance between the two closest groups is entered. The procedure is conducted with reference to the following conditions:
a. Maximum number of steps is 16.
b. Minimum partial F to enter is 3.84.
c. Maximum partial F to remove is 2.71.
d. F level, tolerance, or VIN insufficient for further computation.

The Eigenvalue of the canonical discriminant functions that is used in the context of the analyses following the Mahalanobis distances approach is 0,404. The canonical correlation is with 0,536 very sufficient. Wilk's lambda is highly significant with 0,000. The calculated standardized canonical discriminant function coefficients are 0,420 (age), 0,433 (voting relevance), -0,552 (gender) and 0,383 (information time). The following table shows the unstandardized coefficients:

Table 8: Unstandardized Canonical Discriminant Function Coefficients (MDA)

	Discriminant Function Coefficients
Age	,023
Votingrel	,238
Gender	-1,187
Infotime	,042
(Constant)	-1,519

Although the results of the discriminant function are already of good quality and 74,7 % of the grouped cases are classified correctly, the analysis was also run using the Wilk's lambda approach as the grouping criterion. Indeed, the classification correctness could even be improved. Consequently, the results of this procedure are presented as follows.

However, the interpretation of the results with respect to the different procedure is stated at the end of this chapter including the results of the binary logistic regression. The table below shows the casewise statistics of the first 20 cases including the discriminant score:

Table 9: Casewise Statistics

No.	Act. Gr.	Highest Group					Second Highest Group			Dis. Sc.
		Prd. Grp..	P(D>d | G=g)		P(G=g | D=d)	sMDtC	Gr.	P(G=g | D=d)	sMDtC	
			P	df						
1	1	1	,563	1	,526	,335	0	,474	,546	,107
2	0	1**	,585	1	,537	,299	0	,463	,593	,139
3	1	1	,742	1	,786	,108	0	,214	2,709	1,014
4	0	0	,790	1	,772	,071	1	,228	2,508	-,898
5	0	0	,463	1	,862	,539	1	,138	4,207	-1,37
6	0	0	,472	1	,860	,516	1	,140	4,144	-1,35
7	1	0**	,976	1	,696	,001	1	,304	1,655	-,601
8	0	0	,530	1	,845	,395	1	,155	3,784	-1,26
9	0	1**	,951	1	,687	,004	0	,313	1,577	,624
10	1	0**	,498	1	,853	,459	1	,147	3,979	-1,31
11	1	1	,074	1	,962	3,196	0	,038	9,639	2,473
12	0	0	,707	1	,592	,141	1	,408	,886	-,255
13	1	1	,678	1	,579	,173	0	,421	,813	,270
14	0	0	,549	1	,519	,359	1	,481	,515	-,032
15	0	1**	,635	1	,816	,225	0	,184	3,209	1,160
16	1	0**	,629	1	,557	,234	1	,443	,695	-,148
17	0	0	,709	1	,796	,139	1	,204	2,857	-1,01
18	1	0**	,878	1	,660	,023	1	,340	1,354	-,478

19	0	0	,971	1	,714	,001	1	,286	1,831	-,667
20	1	1	,756	1	,782	,097	0	,218	2,652	,997
**. Misclassified case										

In the case of the discriminant analysis according to the Wilk's lambda approach, the Eigenvalue is 0,439 and slightly better than before. The canonical correlation is 0,552 and Wilk's lambda is again highly significant (0,000).

As this calculation method does not follow a stepwise procedure, all tested parameters are included in the model. The standardized canonical discriminant function coefficients are 0,271 (age), -0,043 (investor type), 0,444 (voting relevance), -0,527 (gender), -0,190 (trading experience), 0,369 (information time), 0,238 (professional status) and -0,047 (educational level). The following table shows the unstandardized coefficients:

Table 10: Unstandardized Canonical Discriminant Function Coefficients (WLA)

	Discriminant Function Coefficients
Age	,015
Invtype	-,066
Votingrel	,244
Gender	-1,135
Tradingexp	-,094
Infotime	,040
Profstatus	,265
Educ	-,050
(Constant)	-,956

The overall grouping quality is very satisfying. 76 % grouped cases are classified correctly. However, this slightly better result has to be regarded in the context that the number of independent variables used in this model has increased in comparison with the stepwise method.

As the binary logistic regression also shows that only these four parameters that are already included in the stepwise method have a significant influence, the hypotheses testing is answered with reference to this output. Nevertheless, the two following figures illustrate the distribution of the values of the canonical discriminant function divided by groups (left: eth./right uneth.):

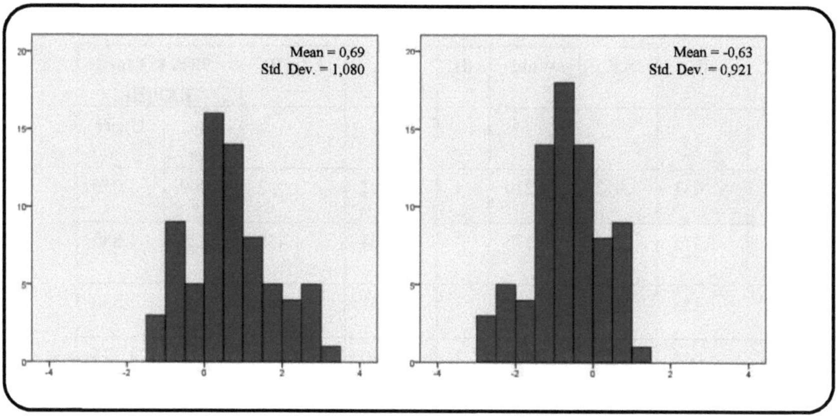

Figure 7: Distribution of Canonical Discriminant Function divided by Groups

Apart from calculating discriminant functions to separate the individuals with reference to their decisions according to their personal attributes, the binary logistic regression facilitate direct conclusions about significant characteristics that makes participants to come to moral choices.

With reference to the model section of this substudy, the impact of eight different parameters was tested. The calculation is terminated at the fourth step as the parameter estimation at iteration five does not vary significantly. The Nagelkerke R Square at the fourth step is 0,392; the Cox & Snell R Square is 0,294. The following table illustrates the variables in the equation after the fourth step:

Table 11: Variables included in binary logistic Regression Analysis

Step 4	**B**	**S.E.**	**Wald**	**df**	**Sig.**	**Exp(B)**	**95% C.I.for EXP(B)**	
							Lo-wer	Upper
Age	,031	,012	7,210	1	,007	1,032	1,009	1,056
Vo-tingrel	,375	,131	8,176	1	,004	1,454	1,125	1,880
Gender	-1,423	,421	11,413	1	,001	,241	,106	,550
Info-time	,064	,027	5,422	1	,020	1,066	1,010	1,125
Consta nt	-2,895	1,276	5,147	1	,023	,055		

In the context of multivariate analyses, multicollinearity has to be considered to avoid the misinterpretation of biased results. The correlations of the predictor variables are regarded as being of an acceptable level and no correlation between the independent variables included in the binary logistic regression model after the fourth step indicates multicollinearity problems.

It can be summarized that four significant determinants drive individuals to make up ethical decision. These are age (0,007), voting relevance (0,004), gender (0,001) and information time (0,020). All parameters show a positive impact which means that a higher level tend to influence the consideration of ethical aspects positively. In the case of the gender, female investors consider moral aspects more comprehensively. Based on the binary logistic regression, 75 % of the decision makers that do not care about ethical issues and 68,6 % of those who care about moral aspects in the financial decision process can be predicted correctly.

As the participants are not splitted equally in both groups in this study, overall 71,9 % are predicted correctly. Consequently, hypotheses 8, 10, 11 and 13 are confirmed; hypotheses 9, 12, 14 and 15 have to be rejected.

5.4.3 Testing of Intention, Changing Probability and Risk Attitude Hypotheses

The impact of the a priori intention to consider moral aspects during the financial decision making process on the average probability to change a decision over all five scenarios with the introduction of risk is tested using a linear regression setting. The personal degree of risk aversion is introduced as a moderating variable in this context. Due to the divergent levels of scale, the regression is conducted on standardized values.

Both tested models, including and excluding the moderating variable, show significant results. However, the tested model including the moderator variable (model 2) shows a higher and very satisfying level of explanatory power. The R Square of this model is 0,568; the adjusted R Square is 0,559. The following table illustrates the most relevant information regarding the coefficients of both models:

Table 12: Model Coefficients including/excluding Moderator

Model		**Unstd. C.**		**Std. C.**	**T**	**Sig.**	**95,0% Conf. Int. for B**		**Collinearity Statistics**	
		B	SE	Beta			Low B.	Up B.	Tol-er-ance	VIF
1	(Con-stant)	3,2	,045		70,4	,000	3,10	3,27		
	Zscore (moral-intent)	-,56	,045	-,688	-12	,000	-,65	-,47	,997	1,003
	Zscore (riska-version)	,26	,045	,319	5,7	,000	,169	,348	,997	1,003
2	(Con-stant)	3,2	,045		71,2	,000	3,10	3,27		
	Zscore (moral-intent)	-,55	,045	-,674	-12	,000	-,64	-,46	,986	1,014
	Zscore (riska-version)	,25	,045	,306	5,51	,000	,159	,337	,987	1,013
	Modera-tor	,10	,046	,127	2,28	,024	,014	,195	,980	1,021

In addition to the variance inflation factors that are shown in the table above, further collinearity diagnostics are calculated in order to evaluate the adequateness of the tested models. The table below provides a more comprehensive overview. It can be concluded that the stability and variance of the regression estimates can be considered as being of sufficient quality.

Table 13: Collinearity Diagnostics

Mod.	Dimension	Eigen-value	Cond. Index	Variance Proportions			
				(Constant)	Zscore (moral-intent)	Zscore (risk-aversion)	Moderator
1	(Constant)	1,052	1,000	,00	,47	,47	
	Zscore (moral-intent)	1,000	1,026	1,00	,00	,00	
	Zscore (riskaversion)	,948	1,054	,00	,53	,53	
2	(Constant)	1,128	1,000	,08	,17	,12	,49
	Zscore (moral-intent)	1,052	1,036	,00	,44	,50	,00
	Zscore (riskaversion)	,994	1,066	,87	,06	,06	,01
	Moderator	,826	1,168	,05	,34	,31	,49

The expectation was that participants with high moral intentions are less willing to change their decisions in contrast to others. As the regression model excluding the moderator variable is significant and the standardized intention coefficient is negative (-12,271; 0,000), hypothesis 16 is confirmed.

In hypothesis 17 it is proposed that this effect is moderated by the degree of personal risk aversion. Broadly said this means that participants with a high level of risk aversion are assumed to realize their ethical intentions less intensive and are more likely to change their decision than those who are risk affine.

As the moderator of the tested second regression is positive (2,287) and the overall model is significant including the moderator variable (0,024), hypothesis 17 is confirmed.

5.5 Discussion and Limitations

The aim of this substudy was to research the relationship between corporate disclosures and the information assessment of the disclosed information by financial investors comprehensively. The results of this substudy suggest that the concept of decision types, personal attributes of the investors and decision making procedures in the light of different dilemmas are closely connected.

It could be tested empirically, that attributes like the age, the relevance of the shareholder's voting right, the gender and the individual time spent on financial news are determinants that drive investors overall to consider moral aspects during their financial decision processes.

Moreover, the personal a priori intention to regard moral issues reduces the willingness to change financial decisions with the exposure to financial risk. However, this relationship is negatively moderated by the individual degree of risk aversion.

With respect to specific dilemma types, it can be summarized that participants who follow the rule utilitarianism or rule deontology approach consider ethical aspects more comprehensively and rule deontologists also tend to be less willing to change their decisions. In the context of other scenarios like the physical environment vignette, the decision type is irrelevant with respect to the willingness to decision changes.

As the first substudy of this work established the link between the quality of corporate disclosures and financial key figures of traded securities, the second substudy support the explanation of this effect with respect to the individual decision making process based on information from corporate disclosures.

This study supports the development of a comprehensive understanding of the relationship between corporate disclosures and the financial risk assessment of investors. It contributes to the development of complex econometrical models in the field of corporate reporting and financial investment that cover individual factors as well as the situation related context of decision routines.

This is the first comprehensive model that combines quantitatively moral decision types, situational dilemmas that arise through information from corporate disclosures and personal attributes like the degree of risk aversion.

In addition to the advancement of theoretical knowledge, the practical pertinence of these results affects several parties that are involved in the area of corporate reporting. Firstly, reporting corporations gain information about the insights of decision processes of investors and the role that information from corporate disclosures plays in this context.

With reference to their business strategy and the structure of their shareholders, they might optimize their reporting behavior as dilemma related as well as characteristic related decisions are partly predictable.

Additionally, the relevance and extent to communicate issues that are closely related to financial risk aspects in the light of divergent risk aversions of the investors might differ with respect to the investors' structure. Secondly, the results of this substudy support investors to understand and objectify their financial decision processes in order to overcome potential biases during their decision routines. Especially, the self-confrontation with the own decision type and its consequences might be helpful in the light of decision dilemmas.

Thirdly, regulatory authorities might also benefit as this substudy supports them to foresee the consequences if they make specific reporting issues (e.g. environmental ad hoc reporting aspects) compulsive with respect to decision processes of investors.

All in all, this second substudy has several limitations that have to be mentioned. The sampling can be criticized as it consists of a group of financially interested private investors that attended a fair. The sample is not only limited to 146 participants but also consists of individuals with comparable or even equal cultural backgrounds.

Moreover, the scenarios that are used in the context of this study are based on the work of Fritzsche and Becker (1984) and cover only a limited number of decision dilemmas. Additionally, these vignettes have not been developed for purposes in the context of financial decision making and corporate disclosures and had to be adjusted. In addition to that, the measurability of several parameters like the degree of risk aversion is difficult and biases might occur.

Lastly, survey related criticism with reference to reliability and validity as well as error rate related issues with respect to the conducted statistical analyses have to be taken into consideration.

6 Synthesis of Developed Models

This chapter provides a simplifying synthesis of the empirical analyses that have been conducted in the scope of this study. The aim of this section is not to give a detailed review of the research results but to present them in their structural context. The following figure serves as the basis of further explanations.

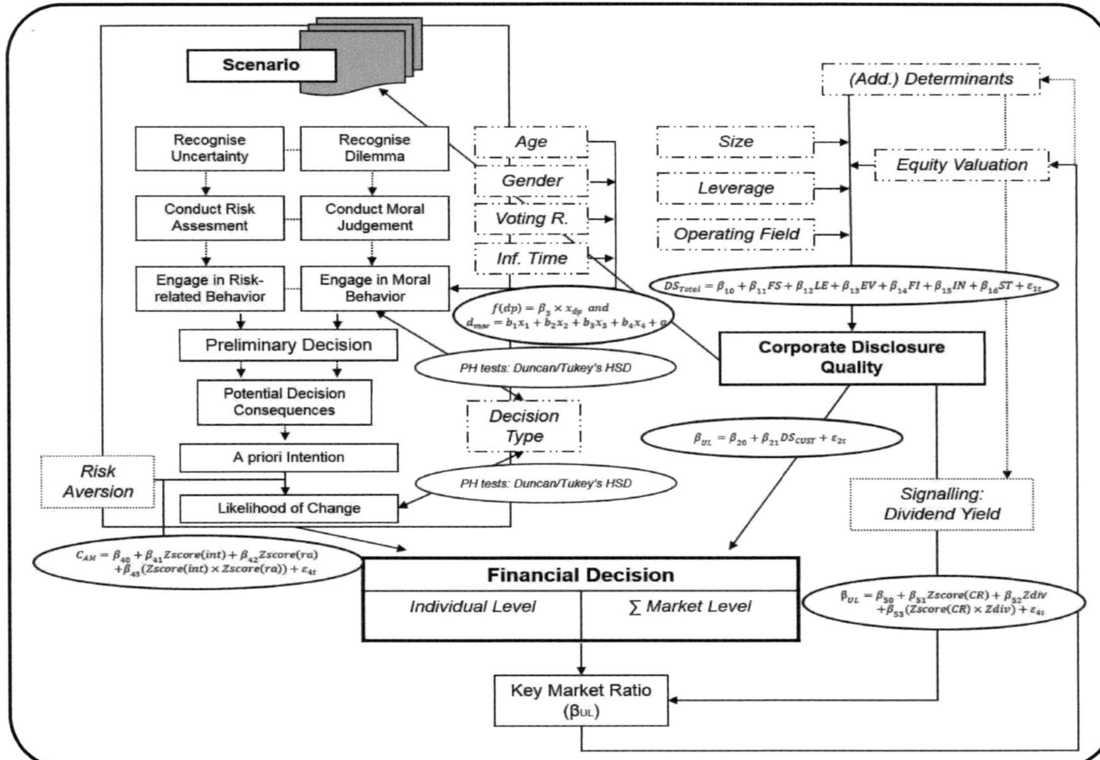

Figure 8: Synthesis of Tested Models (the formulas of this illustration are presented in the corresponding section)

As illustrated in the figure above, the overall objective of this work is to investigate the relationship between corporate disclosures, the information processing of investors with a focus on the risk assessment and the final financial decision as well as its consequences as a complex multivariate process.

Due to technical reasons, the work is divided into two parts. Financial decisions are researched on the microeconomic and the macroeconomic level. A scenario-based study is used to investigate decision processes on the individual level; an approach based on a scoring model is applied to research the overall market effects of corporate disclosures on financial key risk indicators.

The logical first step of this research approach is to proof whether a significant relationship between corporate disclosures and key market data exist. To this end, several determinants have been tested in the context of a multivariate regression model that might drive the disclosure quality of corporations.

As visualized at the top right-hand side of the figure, the disclosure quality is driven by the corporate size, the leverage, the equity valuation, the operating field and additional undefined factors. With the application of a further regression model including a moderator variable, it could be shown that the disclosure quality impacts unlevered beta factors as a systematic risk indicator.

This effect is moderated by the dividend yield which is interpreted as a credibility indicator with reference to the disclosed information. The observed effects build up a reciprocal model as the unlevered betas directly and indirectly influence determinants that drive the disclosure quality. The logical second step is to investigate the individual decision process detailed.

On the basis of the evaluation of corporate disclosures and referring to a theoretical framework, several scenarios have been elaborated to investigate the individual level of financial decision making processes regarding corporate disclosures.

This part of the work is illustrated on the left side of the figure above. A discriminant function could be calculated to estimate whether individuals consider ethical aspects during their overall financial decision process. The determinants could also be confirmed through the application of a binary logistic regression model.

Moreover, the significant association between the personal decision type and scenario related decisions as well as the likelihood of decision changes was tested empirically using analyses of variances and different types of post hoc tests.

Additionally, the exposure of financial risk was simulated and the participants were confronted with potential negative financial consequences related to their decisions. It could be shown that the individual a priori intention impacts the likelihood to change the decision with the introduction of financial risk significantly.

Nevertheless, a regression model including the individual risk aversion as a moderator variable showed that risk averse individuals are more sensitive for a decision change although the a priori intention would not have indicated this.

This study provides insights for both levels: the individual decision process based on corporate disclosures and the influence of corporate disclosures on relevant market data. Indeed, the relationship is very clear. The market decision and finally the market data represents the sum of individual decisions which is illustrated with the bold-framed box at the bottom in the figure above. Apart from institutional investors that have not been researched explicitly in the context of this work, the results of the empirical analyses provide a very comprehensive overview of the research field.

Consequently, the synthesis of the tested models of this study supports the overall understanding of the role of corporate disclosures in the framework of asymmetric information with reference to the financial risk assessment.

7 Synopsis

The objective of this work was to connect corporate disclosures and the financial risk assessment of investors through quantitative models in order to elaborate observations that add new academic insights and practical, pertinent knowledge.

The application of multivariate scoring models and statistical analyses showed an innovative link between risk related market indicators and the quality of corporate disclosures. This relationship is characterized by the influence of credibility, namely the dividend yield as a costly corporate plausibility indicator. The validation tests of quality determinants converge intensively to previous research results and support clearly the adequateness of the tested models.

In addition to the market data level, individual decision procedures have been researched in order to break down the relationship between corporate disclosures and the financial risk assessment to the most detailed level.

The objective of this step was to capture behavioural and cognitive activities during decision routines with reference to investments in financial securities based on disclosure related information. The application of established scenario techniques and self-developed and empirically proven econometrical models showed that on this microeconomic level individual characteristics impact financial decisions significantly.

Additionally, it could be shown that decision types have an outstanding relevance over varying scenarios also with regard to the exposure to financial risk. The integration of risk aversion in this context make this study unique as it was tested for the first time that the realization of the a priori intention to make up financial decisions is moderated by the risk attitude with reference to information from corporate disclosures.

As established in the synthesis section, the combination of the empirical results of the individual and market level lead to new insights and a comprehensive understanding: Individuals consider information from corporate disclosures during their financial decision making processes. The impact varies with the decision type, scenario related factors and further individual characteristics.

On the market level, corporations can improve risk indicators through the disclosure of comprehensive information. Plausibility signals improve the effectiveness of the reduction of information asymmetries.

The practical pertinence of these results is multi-dimensional. The reduction of corporate beta factors as tested in the context of the first substudy means that investors reward the reduction of information asymmetries.

Practically speaking, companies can reduce their capital costs through the publication of certain information. Investors honour the disclosure of customer related information published by service and trade companies extraordinarily. With regard to corporate dividend policies, the positive influence of the dividend yield as a moderating variable in this context should be considered and improves the positive effect that arises due to the reduction of information asymmetries.

Moreover, companies that know about the structure of their shareholders and make use of the results of the second substudy might also be able to assist them during financial procedures and convince them to invest in their securities.

On the other hand, the results also support investors to objectify their decision processes with the knowledge of potential biases and the influence of personal characteristics. Lastly, state authorities that regulate the degree of disclosure obligations might be inspired to recheck the requirements in order to reach a higher level of market efficiency in practice.

Appendix

Appendix A

Variable Measurement of first Substudy

Variable	Measurement
Beta	Corporate beta factors of the year 2011 (source: Deutsche Boerse)
Beta (unlevered)	Calculated based on capital structure & tax rate by using CAPM assumptions
Customer-related Score	subscore of the total sustainability score
Debt	total debt in 2010 (source: Datastream, code: WC03255)
Dividend Yield	dividend per share as a percentage of the share price by the end of 2010 (source: Datastream, code: DY)
Equity (market value)	share price multiplied by the number of ordinary shares in issue by the end of 2010 (source: Datastream, code: MV)
Equity (book value)	book value of equity in 2010 (source: Datastream, code: WC03501)
Equity Valuation	equity (market value) divided by equity (book value)
Leverage	total debt of total capital in 2010 (source: Datastream, code: WC08221)
Operating Field	3 groups: Finance, Industry, Service&Trade (dummy variables)
Size	nat. log. of total assets (book value) of the corporation in 2010 (source: Datastream, code: WC02999)
Sustainability Score	disclosures of the business year 2010 are analyzed by using analytical grids, quantitative scores are determined in relation to the quality of information
Tax rate	consists of uniform national corporate tax rate plus solidarity surcharge and average local tax rate in 2011 (source: Federal Statistical Office of Germany)

Appendix B

Datastream Company Codes of Analyzed Corporations

D:ARL,	D:NDA,	D:SPR,	D:BYW6,	D:GBF,	D:BNR,
D:CLS1,	D:CON,	D:D9CX,	D:DEQ,	D:DWNI,	D:DOU,
F:EADS,	D:ZIL2,	D:FIE,	D:FRA,	D:FPE,	D:GFJ,
D:G1A,	D:GXI,	D:GIL,	D:HHFA,	D:HNR1,	D:HDD,
D:HOT,	D:BOSS,	D:IVG,	D:KD8,	D:KCO,	D:KRN,
D:LXS,	D:LEO,	D:MTXX,	D:PRA,	D:PSMX,	D:PUM,
D:RAA,	D:RHM,	D:RHK,	D:SZG,	D:SGL,	D:PRE,
D:SAZ,	D:SZU,	D:SY1,	D:TGM,	D:TUI1,	D:VOS,
D:WCH,	D:WIN				

Appendix C

Grid of Supplier-related Disclosure Analysis

No.	ITEM	Observation	PTS	Observation	PTS	Observation	PTS	Dimension
1	Positioning related to business strategy (CEO`s editorial or comparable)	No information provided	1	Simple reference to relevance of suppliers	5	Importance mentioned and basic additional information provided	9	-
2	Relevance within reporting framework (annual or other report)	No supplier related headings or chapters	1	Supplier related heading(s)	5	Supplier related chapter heading	9	-
3	Type of collaboration	No information provided	1	Types of collaboration are named (Contractual, Joint-Ventures etc.)	5	Collaboration types are named and future perspectives are described	9	Economic

4	Finan-cial Com-pensa-tion	No infor-mation provided	1	Type of compen-sation is named (e.g. per-for-mance-oriented)	5	Detailed compensa-tion-related information or future perspec-tives are described	9	Economic
5	Selec-tion cri-teria and anti-corrup-tion	No infor-mation provided	1	Selection criteria and ne-cessity of social standards are men-tioned	5	Country of origin (CoO) with high social standards or contrac-tual mini-mum social require-ments are named and anti-corruption guidelines	9	Social
6	Proof of compli-ance with so-cial stand-ards	No infor-mation provided	1	Compli-ance with own so-cial standards was test-ed initial-ly	5	Compli-ance and regular tests or CoO with high social standards	9	Social
7	Envi-ron-mental selec-tion cri-teria	No infor-mation provided	1	Necessity of envi-ronmen-tal stand-ards are men-tioned	5	Binding environ-mental re-quirements or ISO/DIN certifica-tion named	9	Environ-mental

8	Pollution Treatment	No information provided	1	Single aspects like resource consumption or site renovation are named	5	Environmental pollution of total production process is measured	9	Environmental

minimum score: 8; maximum score: 72

Appendix D

Grid of Employee-related Disclosure Analysis

No.	ITEM	Observation	PTS	Observation	PTS	Observation	PTS	Dimension
1	Positioning related to business strategy (CEO`s editorial or comparable)	No information provided	1	Simple reference to relevance of employees	5	Importance mentioned and basic additional information provided	9	-
2	Relevance within reporting framework	No employee related headings or chapters	1	Employee related heading(s)	5	Employee related chapter heading	9	-
3	Financial compensation of employees (not management & supervisory board)	No information provided	1	General information with reference to performance and long-term orientation	5	Incentives & perspectives regarding e.g. employee shares/pension model	9	Economic
4	Structural distribution of paid financial compensation or comparable	No information provided	1	Simple structure according to business units or comparable	5	Innovation-oriented structuring (e.g. administration, R&D, sales)	9	Economic

5	Training and education efforts	No information provided	1	Basic information on successfully completed trainings are provided	5	Detailed information about content or duration and taught group of people	9	Social
6	Recruiting and employee diversity (not management & supervisory board)	No information provided	1	Information on employee diversity are provided	5	Activities (esp. recruiting) to strengthen employee diversity	9	Social
7	Sensitizing for environmental issues in production process	No information provided	1	Basic information on activities to reduce resource consumption are provided	5	Organized training or activities to reduce unnecessary consumption or certification or detailed report on environmental figures	9	Environmental
8	Dealing with employee feedback on environmental or innovative issues	No information provided	1	Feedback is systematically collected	5	Incentives for feedback provision or systematic ideas management	9	Environmental

minimum score: 8; maximum score: 72

Appendix E

Grid of Customer-related Disclosure Analysis

No.	ITEM	Observation	PTS	Observation	PTS	Observation	PTS	Dimension
1	Positioning related to business strategy (CEO`s editorial or comparable)	No information provided	1	Simple reference to extraordinary relevance of customers	5	Importance mentioned and basic additional information provided	9	-
2	Relevance within reporting framework	No customer related headings or chapters	1	Customer related heading(s)	5	Customer related chapter heading	9	-
3	Sales and marketing activities	No information provided	1	General information regarding use of distribution channels	5	Relevance of distribution channels in future and perspectives	9	Economic
4	Customer satisfaction	No information provided	1	Customer satisfaction, loyalty or comparable is measured	5	Activities to boost customer satisfaction or loyalty are named	9	Economic
5	Social activities in favour of (future) customers	No information provided	1	Social responsibility is mentioned	5	Concrete activities to support cultural or social integration	9	Social

6	Price policy / Product Development	No information provided	1	Customer structure and / or its possible development is reported	5	Price differentiation or product development according to social aspects is described	9	Social
7	Environmental product / service characteristics	No information provided	1	Recycling or other sustainable approaches	5	Detailed information on recycling and customers are provided with information for environmentally friendly product use or comparable	9	Environmental
8	Dealing with customer feedback on environmental or innovative issues	No information provided	1	Feedback is systematically collected	5	Incentives for feedback provision or systematic analysis of feedback exist	9	Environmental

minimum score: 8; maximum score: 72

Appendix F

Variable Measurement of second Substudy

Variable	Measurement
Age	age
Avgchange	arithmetic mean of willingness to change decision over all scenarios
Educ	educational level: schooling, high school diploma, professional degree, university degree
Gender	female, male
Infotime	time spent on financial/economic information per day in minutes
Invtype	self-assessment of investor-type: growth-investor, value-investor, other
Moderator	mathematical product of the standardized intention and the standardized degree of risk aversion
Profstatus	professional status: employed for wages, self-employed, retired, other
Tradingexp	trading/financial experience, coded with a scale of 10
Votingrel	relevance of voting right as a shareholder, coded with a scale of 10
Zscore(moralintent)	standardized a priori moral intention regarding financial decisions
Zscore(riskaversion)	standardized degree of individual risk aversion

Bibliography

Aerts, W., Cormier, D. and Magnan, M. (2006). Intra-industry imitation in corporate environmental reporting: An international perspective, Journal of Accounting and Public Policy, 25(3), 299-331.

Aerts, W., Cormier, D. and Magnan, M. (2007). The association between web-based corporate performance disclosure and financial analyst behaviour under different governance regimes, Corporate Governance: An International Review, 15(6), 1301–1329.

Akerlof, G. (1970). The market for 'Lemons': quality uncertainty and the market mechanism, Quarterly Journal of Economics, 84(3), 488-500.

Allen, F., Bernardo, A. and Welch, I. (2000). A theory of dividends based on tax clienteles, Journal of Finance, 55(6), 2499-2536.

Andriof, J. and Waddock, S. (2002). Unfolding stakeholder engagement, In Andriof, J., Waddock, S., Husted, B. and Rahman. S. (Eds.), Unfolding stakeholder thinking. theory, responsibility and engagement, Sheffield: Greenleaf.

Archel, P., Husillos, J., Larrinaga, C., Spence C. (2009). Social disclosure, legitimacy theory and the role of the state, Accounting, Auditing and Accountability Journal, 22(8), 1284–1307.

Baret, P. (2006). L'évaluation contingente de la performance globale des entreprises: une méthode pour fonder un management socialement responsable?, In Rosé, J.-J. (Ed.), Responsabilité sociale de l'entreprise, méthodes et recherches management, Brussels: DeBoeck.

Barthes, R. (1981). Introduction à l'analyse structurale des récits, Paris, Le Seuil, Collection Points Essais.

Belkaoui, A., Karpik, P. G. (1989). Determinants of the corporate decision to disclose social information, Accounting, Auditing and Accountability Journal, 2(1), 36-51.

Bhushan, R. (1989). Firm characteristics and analyst following, Journal of Accounting and Economics, 11(2), 255-274.

Boesso G. and Kumar, K. (2007). Drivers of corporate voluntary disclosure: A framework and empirical evidence from Italy and the United States, Accounting, Auditing and Accountability Journal, 20(2), 269-296.

Botosan, C. A. (1997). Disclosure level and the cost of equity capital, The Accounting Review, 72(3), 323-350.

Botosan, C. A. (2006). Disclosure and the cost of capital: What do we know?, Accounting and Business Research, 36, 31-40.

Botosan, C. A. and Plumlee, M. A. (2002). A re-examination of disclosure level and the expected cost of equity capital, Journal of Accounting Research, 40(1), 21-40.

Brown, L. D. and Higgins, H. N. (2001). Managing earnings surprises in the US versus 12 other countries, Journal of Accounting and Public Policy, 20(4-5), 373–398.

Buzby, S. (1975). Company size, listed versus unlisted stocks, and the extent of financial disclosure, Journal of Accounting Research, 13(1), 16-37.

Campbell, J. Y., Lettau, M., Malkiel, B. G. and Xu, Y. (2001). Have individual stocks become more volatile? An empirical exploration of idiosyncratic risk, Journal of Finance, 56(1), 1-43.

Caroll, A. (1979). A three-dimensional conceptual model of corporate social performance, Academy of Management Review, 4(4), 497-505.

Casson, M. (1996). The comparative organisation of large and small firms: An information cost approach, Small Business Economics, 8(5), 329-345.

Chen, S. and Bouvain, P. (2009). Is corporate responsibility converging? A comparison of corporate responsibility reporting in the USA, UK, Australia and Germany, Journal of Business Ethics, 87(1), 299-317.

Cheng, Q. and Lo, K. (2006). Insider trading and voluntary disclosures, Journal of Accounting Research, 44(5), 815–848.

Clarkson, M. E. (1995). A stakeholder framework for analyzing and evaluating corporate social performance, Academy of Management Review, 20(1), 92-117.

Clarkson P., Guedes J. and Thompson, R. (1996). On the diversification, observability, and measurement of estimation risk, The Journal of Financial and Quantitative Analysis, 31(1), 69-84.

Clarkson, P., Kao, J. and Richardson, G. (1999). Evidence that management discussion and analysis (MD&A) is a part of a firm's overall disclosure package, Contemporary Accounting Research, 16(1), 111-134.

Cochran, P. L. and Wood, R. A. (1984). Corporate social responsibility and financial performance, Academy of Management Journal, 27(1), 42-56.

Cormier, D., Aerts, W., Ledoux, M. J. and Magnan, M. (2009). Attributes of social and human capital disclosure and information asymmetry between managers and investors, Canadian Journal of Administrative Sciences, 26(1), 71-88.

Cormier, D., Ledoux, M. J. and Magnan, M. (2011). The informational contribution of social and environmental disclosures for investors, Management Decision, 49(8), 1276-1304.

Cormier, D., Ledoux, M. J. and Magnan, M. (2012). The moderating effect of voluntary disclosure on the relation between earnings quality and information asymmetry: some Canadian evidence, International Journal of Accounting, Auditing and Performance Evaluation, 8(2), 157-183.

Cormier, D. and Magnan, M. (2002). Performance reporting by oil and gas firms: contractual and value implications, Journal of International Accounting, Auditing and Taxation, 11(2), 131–153.

Cormier, D. and Magnan, M. (2003). Environmental reporting management: a continental European perspective, Journal of Accounting and Public Policy, 22, 43–62.

Cormier, D. and Magnan, M. (2007). The revisited contribution of environmental reporting to investors' valuation of a firm's earnings: An international perspective, Ecological Economics, 62(3-4), 613-626.

Cormier, D., Magnan, M. and van Velthoven, B. (2005). Environmental disclosure quality in large German companies: Economic incentives, public pressures or institutional conditions?, European Accounting Review, 14(1), 3-39.

Cowles, A. and Jones, H. E. (1937). Some a posteriori probabilities in stock market action, Econometrica, 5(3), 280-294.

Crawford, V. (1998). A survey of experiments on communication via cheap talk, Journal of Economic Theory, 78(2), 286-298.

Cronqvist, H. and Nilsson, M. (2005). The choice between rights offerings and private equity placements, Journal of Financial Economics, 78(2), 375–407.

Dahlman, C. J. (1979). The problem of externality, Journal of Law and Economics, 22(1), 141–162.

Déjean, F. and Martinez, I. (2009). Environmental disclosure and the cost of equity: the French case, Accounting in Europe, 6(1), 57-80.

Demazière, D. and Dubar, C. (1997). Analyser les entretiens biographiques, Paris, Essais et recherches Nathan.

Depoers, F. (2000). A cost benefit study of voluntary disclosure: some empirical evidence from French listed companies, European Accounting Review, 9(2), 245-263.

Dhaliwal, D. S., Radhakrishnan, S., Tsang, A. and Yang, Y. G. (2012). Nonfinancial disclosure and analyst forecast accuracy: International evidence on corporate social responsibility disclosure, The Accounting Review, 87(3), 723-759.

Dhaliwal, D. S., Zhen Li, O., Tsang, A. and Yang Y. G. (2011) Voluntary nonfinancial disclosure and the cost of equity capital: the initiation of corporate social responsibility reporting, The Accounting Review, 86(1), 59-100.

Diamond, P. A. and Stiglitz, J. E. (1974). Increases in risk and in risk aversion, Journal of Economic Theory, 8(3), 337–360.

Diamond, D. W. and Verrecchia, R. E. (1991). Disclosure, liquidity and the cost of capital, The Journal of Finance, 46(4), 1325-1359.

Dye, R. (1985). Disclosure of nonproprietary information, Journal of Accounting Research, 23(1), 123-145.

Eckbo, B. E. and Verma, S. (1994). Managerial shareownership, voting power and cash dividend policy, Journal of Corporate Finance, 1, 33-61.

Eisenhardt, K. E. (1989). Agency theory: an assessment and review, The Academy of Management Review, 14(1), 57-74.

Entwistle, G. (1999). Exploring the R & D disclosure environment, Accounting Horizons, 13(4), 323-341.

Fama, E. (1970). Efficient capital markets, a review of theory and empirical work, Journal of Finance, 25(2), 383-417.

Fama, E. and French, K. (1992). The cross-section of expected stock returns, Journal of Finance, 47(2), 427-465.

Fama, E. and French, K. (2002). Value versus growth: the international evidence, Journal of Finance, 53(6), 1975–1999.

Ferrell, O. C. and Gresham, L. (1985). A contingency framework for understanding ethical decision making in marketing, Journal of Marketing, 49(3), 87–96.

Fraedrich, J. P. and Ferrell, O. C. (1992). The impact of perceived risk and moral philosophy type on ethical decision making in business organizations, Journal of Business Research, 24(4), 283-295.

Freeman, R. E. and Reed, D. L. (1983). Stockholders and stakeholders: a new perspective on corporate governance, California Management Review, 25(3), 88-106.

Freeman, R. E. (1984). Strategic management: A stakeholder approach, Series in Business and Public Policy, Boston: Pitman Publishing.

Freeman, R. E. (1994). The politics of stakeholder theory: some future directions, Business Ethics Quarterly, 4(4), 409-421.

Friedman, M. (1962). Capitalism and freedom, Chicago: University of Chicago Press.

Fritzsche, D. J. and Becker, H. (1984). Linking management behavior to ethical philosophy - an empirical investigation, Academy of Management Journal, 4(1), 166–175.

Ghazali N. A. M. (2007). Ownership structure and corporate social responsibility disclosure: some Malaysian evidence, Corporate Governance, 7(3), 251-266.

Gibson, K. (2003) Contrasting role morality and professional morality: implications for practice, Journal of Applied Philosophy, 20(1), 17–29.

Giffen, R. (1891). The Gresham Law, The Economic Journal, 1(2), 304-306.

Glover, S. H., Bumpus, M. A., Logan, J. E. and Ciesla, J. R. (1997). Re-examining the influence of individual values on ethical decision making, From the Universities to the Marketplace: The Business Ethics Journey, 16(12-13), 109-119.

Goodpaster, K. (1991). Business ethics and stakeholder analysis. Business Ethics Quarterly, 1(1), 53-73.

Goyal, A. and Santa-Clara, P. (2003). Idiosyncratic risk matters!, Journal of Finance, 58(3), 975-1007.

Graham, J. R., Harvey, C. R. and Rajgopal, S. (2005). The economic implications of corporate financial reporting, Journal of Accounting and Economics, 40(1-3), 3-73.

Gray, E. R. and Balmer, J. M. T. (1998). Managing corporate image and corporate reputation, Long Range Planning, 31(5), 695-702.

Grenadier, S. R. and Wang, N. (2005). Investment timing, agency and information, Journal of Financial Economics, 75(3), 493-533.

Guthrie, J. and Parker, L. D. (1989). Corporate social reporting: a rebuttal of legitimacy theory, Accounting and Business Research, 19(76), 343-352.

Hail, L. (2002). The impact of voluntary corporate disclosure on the ex-ante cost of capital for Swiss firms, The European Accounting Review, 11(4), 741-773.

Hannah, S. T., Avolio, B. J. and May, D. R. (2011). Moral maturation and moral conation: a capacity approach to explaining moral thought and action, Academy of Management Review, 36(4), 663-685.

Hayes, R. M. and Lundholm, R. (1996). Segment reporting to the capital market in the presence of a competitor, Journal of Accounting Research, 34(2), 261-279.

Healy, P. M. and Palepu, K. G. (2001). Information asymmetry, corporate disclosure, and the capital markets: a review of the empirical disclosure literature, Journal of Accounting and Economics, 31(1-3), 405–440.

Healy, P., Hutton, A. and Palepu, K. (1999). Stock performance and intermediation changes surrounding sustained increases in disclosure. Contemporary Accounting Research, 16(3), 485-520.

Hillman, A. J. and Keim, G. D. (2001). Shareholder value, stakeholder management, and social issues: what's the bottom line?, Strategic Management Journal, 22(2), 125-139.

Hiltebeitel, K. M. and Jones, S. K. (1992). An assessment of ethics instruction in accounting education, Journal of Business Ethics, 11(1), 37-46.

Hirst, D. E., Koonce, L. and Venkataraman, S. (2008). Management Earnings forecasts: A review and framework, Accounting Horizons, 22(3), 315-338.

Hogner, R. H. (1982). Corporate social reporting: eight decades of development at US steel, Research in Corporate Social Performance and Policy, 4(1), 243-250.

Hollander, S., Pronk, M. and Roelsofsen, E. (2010). Does silence speak? An empirical analysis of disclosure choices during conference calls, Journal of Accounting Research, 48(3), 531-563.

Holthausen, R. and Leftwich, R. (1983). The economic consequences of accounting choice, Journal of Accounting and Economics, 5(1), 77-117.

Husser, J., André, J.-M., Barbat, G. and Lespinet-Najib, V. (2012). CSR and sustainable development: are the concepts compatible?, Management of Environmental Quality, 23(6), 658-672.

Husser, J. and Evraert-Bardinet, F. (2014). The effect of social and environmental disclosure on companies' market value, Management International, 19(1), 61-84.

Husser, J., Gautier, L., André, J.-M. and Lespinet-Najib, V. (2013). Linking purchasing to ethical decision-making: An empirical investigation, Journal of Business Ethics, 123(2), 327-338.

Ittner C. D. and Larcker, D. F. (1998). Are nonfinancial measures leading indicators of financial performance? An analysis of customer satisfaction, Journal of Accounting Research, 36(Suppl.), 1-35.

Jahdi, K. and Acikdilli, G. (2009). Marketing communications and corporate social responsibility: marriage of convenience or shotgun wedding? Journal of Business Ethics, 88(1), 103-113.

Jensen, M. C. (1983). Organization theory and methodology, Accounting Review, 58(2), 319-339.

Jensen, M. C. (1986). Agency costs of free cash flow, corporate finance and takeovers, American Economic Review, 76(2), 323–339.

Jensen, M. C. and Meckling, W. H. (1976). Theory of the firm: managerial behavior, agency costs and ownership structure, Journal of Financial Economics, 3(4), 305-360.

Jensen, M. C. and Meckling, W. H. (1978). Can the corporation survive?, Financial Analysts Journal, 31(1), 31-37.

John, K, and J. Williams. (1985). Dividends, dilution, and taxes: a signaling equilibrium, Journal of Finance, 40(4), 1053-1070.

Jones, D. A. (2007). Voluntary disclosure in R&D-intensive industries, Contemporary Accounting Research, 24(2), 489–522.

Kant, I. (1972). Fundamental principles of the metaphysic of morals, in problems of moral philosophy: an introduction to ethics, Encino (CA): Dickenson.

Kasznik, R. (1999). On the association between voluntary disclosure and earnings management, Journal of Accounting Research, 37(1), 57-81.

Khan, A. M. (1986). Conformity with large speculators: A test of efficiency in the grain futures market, Atlantic Economic Journal, 14(3), 51-55.

Kochan, T.A., Rubinstein, S.A. (2000). Towards a stakeholder theory of the firm: The Saturn partnership, Organization Science, 11(4), 367-386.

Lakhal, F. (2005). Voluntary earnings disclosures and corporate governance: evidence from France, Review of Accounting and Finance, 4(3), 64-85.

Lang, M. and Lundholm, R. (1993). Cross-sectional determinants of analyst ratings of corporate disclosures, Journal of Accounting Research, 31(2), 246-271.

Lang, M. and Lundholm, R. (2000). Voluntary disclosure and equity offerings: reducing information asymmetry or hyping the stock?, Contemporary Accounting Research, 17(4), 623–662.

Lang, M., Lins, K. and Miller, D. (2003). ADRs, analysts, and accuracy: Does cross listing in the United States improve a firm's information environment and increase market value?, Journal of Accounting Research, 41(2), 317-345.

Leland, H. E. and Toft, K. B. (1996). Optimal Capital Structure, Endogenous Bankruptcy, and the Term Structure of Credit Spreads, Journal of Finance, 51(3), 987–1019.

Lemmink J., Schuijf A. and Streukens S. (2003). The role of corporate image and company employment image in explaining application intentions, Journal of Economic Psychology, 24(1), 1-15.

Leuz, C. and Verrecchia, R. E. (2000). The economic consequences of increased disclosure. Journal of Accounting Research, 38 (Suppl.), 91-124.

Lougee, B. and Wallace, J. (2008). The corporate social responsibility trend, Journal of Applied Corporate Finance, 20(1), 96-108.

Luo, X., Wang, H., Raithel, S. and Zheng, Q. (2015). Corporate social performance, analyst stock recommendations, and firm future returns, Strategic Management Journal, 36(1), 123-136.

MacCrimmon, K. R., Wehrung, D. A. and Stanbury, W. T. (1986). Taking risks: the management of uncertainty, New York: The Free Press.

May D. R. and Pauli, K. P. (2002). The role of moral intensity in ethical decision making - a review and investigation of moral recognition, evaluation and intention, Business & Society, 41(1), 84-117.

Markowitz, H.M. (1959). Portfolio selection: efficient diversification of investments, New York: John Wiley & Sons.

McGuire, J. B., Sundgren, A. and Schneeweis, T. (1988). Corporate social responsibility and firm financial performance, Academy of Management Journal, 31(4), 854-872.

McGuire, J. W. (1963). Business and Society, New York: McGraw-Hill.

McMahon, J. M. and Harvey, R. J. (2007). The effect of moral intensity on ethical judgment, Journal of Business Ethics, 72(4), 335-357.

Mercer, M. (2005). The fleeting effects of disclosure forthcomingness on management's reporting credibility. The Accounting Review, 80(2), 723-744.

Miller, M., and K. Rock (1985). Dividend policy under asymmetric information. Journal of Finance, 40(4), 1031-1051.

Modigliani, F. and Miller, M. H. (1958). The cost of capital, corporation finance and the theory of investment, The American Economic Review, 48(3), 261-297.

Modigliani, F. and Miller, M. H. (1963). Corporate income taxes and the cost of capital: a correction. American Economic Review, 53(3), 433–443.

Nagar, V., Nanda, D. and Wysocki, P. (2003). Discretionary disclosure and stock-based incentives, Journal of Accounting and Economics, 34(1-3), 283-309.

O'Fallon, M. J. and Butterfield, K. D. (2005). A review of the empirical ethical decision-making literature: 1996-2003, Journal of Business Ethics, 59(4), 375–413.

Ohlson, J. A. (1995). Earnings, book values and dividends in equity valuation, Contemporary Accounting Research, 11(2), 661-687.

Orlitzky, M., Schmidt, F. L. and Rynes, S .L. (2003). Corporate social and financial performance: a meta-analysis, Organization Studies, 24(3), 403-441.

Pass, C. (2003). Long-term incentive schemes, executive remuneration and corporate performance: an empirical study, Corporate Governance, 3(4), 18-27.

Patten, D. M. (1991). Exposure, legitimacy, and social disclosure, Journal of Accounting and Public Policy, 10(4), 297–308.

Perrini, F. (2005). Building a European portrait of corporate social responsibility reporting, European Management Journal, 23(6), 611–627.

Perrow, C. (1986). Complex organizations. New York: Random House.

Peterson, D., Rhoads, A. and Vaught, B. C. (2001). Ethical beliefs of business professionals: a study of gender, age and external factors, Journal of Business Ethics, 31(3), 225-232.

Phillips, R. (2003). Stakeholder legitimacy, Business Ethics Quarterly, 13(1), 25-41.

Piechocki, R. (2004). Transparency of annual sustainability reports, Corporate Reputation Review, 7(2), 107–123.

Plumlee, M., Brown, D., Hayes, R.M. and Marshall, S. (2010). Voluntary environmental disclosure quality and firm value: further evidence, Social Science Research Network, December 2010, 1-43.

Rettig, S. and Rawson, H. E. (1963). The risk hypothesis in predictive judgments of unethical behavior, Journal of Abnormal and Social Psychology, 66(3), 243-248.

Rogers, J. L. and Stocken, P. C. (2005) Credibility of management forecasts, The Accounting Review, 80(4), 1233-1260.

Ross, S. (1976). The arbitrage theory of capital asset pricing, Journal of Economic Theory, 13(3), 341-360.

Rozeff, M. S. (1982). Growth, beta and agency costs as determinants of dividend payout ratios, Journal of Financial Research, 5(3), 249-259.

Ruckelshaus, W. D. (1989). Toward a sustainable world, Scientific American, 261(3), 166-174.

Russo, M. V. and Fouts, P. A. (1997). A resource-based perspective on corporate environmental performance and profitability, Academy of Management Journal, 40(3), 534–559.

Sharpe, W. F. (1964). Capital asset prices: a theory of market equilibrium under conditions of risk, Journal of Finance, 19(3), 425-442.

Shocker, A. D. and Sethi, S. P. (1973). An approach to incorporating societal preferences in developing corporate action strategies, California Management Review, 15(4), 97-105.

Skinner, D. J. (1995). Do the SEC's safe harbor provisions encourage forward-looking disclosures?, Financial Analysts Journal, 51(4), 38-44.

Tilt, C. A. (1994). The influence of external pressure groups on corporate social disclosure: some empirical evidence, Accounting, Auditing and Accountability Journal, 7(4), 47-72.

Ullmann, A. A. (1985). Data in search of a theory: a critical examination of the relationships among social performance, social disclosure, and economic performance of U.S. firms, Academy of Management Review, 10(3), 540-557.

Verrecchia, R. E. (1983). Discretionary disclosure, Journal of Accounting and Economics, 5, 179-194.

Verrecchia, R. E. (1990). Information quality and discretionary disclosure, Journal of Accounting and Economics, 12(4), 365-380.

Verrecchia, R. E. (2001). Essays on disclosure, Journal of Accounting and Economics, 32(1–3), 97–180.

Watts, R. and Zimmerman, J. (1978). Towards a positive theory of the determination of accounting standards, Accounting Review, 53(1), 112-34.

Watts, R. L. and Zimmerman, J. L. (1990). Positive accounting theory: a ten year perspective, The Accounting Review, 65(1), 131-156.

Webb, K. A., Cahan, S. F. and Sun, J. (2008). The effect of globalization and legal environment on voluntary disclosure, The International Journal of Accounting, 43(3), 219–245.

Williams, J. (1987). Perquisites, risk, and capital structure, Journal of Finance, 42(1), 29–49.

Williamson, O. E. (1985). The economic institutions of capitalism, New York: MacMillian.

Wilmshurst, T. D. and Frost, G. F. (2000). Corporate environmental reporting: A test of legitimacy theory, Accounting, Auditing and Accountability Journal, 13(1), 10-26.

Wolfe, R. A. and Putler, D. S. (2002). How tight are the ties that bind stakeholder groups?, Organization Science, 13(1), 64-80.

Zeghal D. and Ahmed S. A. (1990). Comparison of social responsibility information disclosure media used by Canadian firms, Accounting, Auditing and Accountability Journal, 3(1), 38–53.

Zimmerman, J. L. and Krauss H. H. (1971). Source and magnitude of censure in predictions of unethical behavior, Psychological Reports, 28(3), 727-732.